Mastering Prompts: The Art and Science of Prompt Engineering

Table of Content

1. Introduction to Prompt Engineering: Understanding the Power of Well-Crafted Prompts..................5

2. The Role of Prompts in Natural Language Processing: A Comprehensive Overview..................35

3. Designing Effective Prompts: Principles and Strategies for Maximum Impact..................72

4. Unleashing Creativity: Prompt Variations and Techniques for Generating Diverse Outputs..................74

5. Optimizing Prompt Structures: Crafting Clear, Specific, and Actionable Instructions..................76

6. Leveraging Context: Incorporating Contextual Information in Prompts for Improved Performance..................78

7. Pitfalls and Challenges in Prompt Engineering: Common Mistakes and How to Avoid Them..................80

8. Fine-tuning Models: Techniques and Best Practices for Tailoring Models to Specific Prompts..................82

9. Evaluating Prompt Effectiveness: Metrics and Methods for Assessing

Prompt Performance..84

10. Advanced Prompt Engineering: Incorporating Multimodal Inputs and

Domain-Specific Knowledge...86

11. Ethical Considerations in Prompt Engineering: Addressing Bias,

Fairness, and Responsible AI...88

12. Future Trends and Directions in Prompt Engineering: Shaping the

Path Ahead...90

Chapter 1: Introduction to Prompt Engineering: Understanding the Power of Well-Crafted Prompts

1.1 The Importance of Prompts in Natural Language Processing

Prompts can be defined as explicit instructions or input provided to an AI language model to guide its output generation. They serve as cues or signals that direct the language model's attention and influence the nature of the generated response. The role of prompts in driving AI-generated outputs is significant and multifaceted:

1. **Instructional Guidance: Prompts** act as explicit instructions or guidelines for the language model, shaping the desired type of response. They provide specific information or constraints that steer the model's understanding and generation process.
2. **Conditioning and Priming:** Prompts condition the language model by setting a context or expectation for the generated output. They prime the model's internal state, influencing its interpretation of subsequent text and guiding the generation process accordingly.
3. **Contextual Understanding**: Prompts help language models understand and generate responses that are contextually relevant. By incorporating contextual cues in the prompts, such as previous conversation history or relevant information, models can produce more context-aware and coherent outputs.
4. **Creative Expression:** Prompts play a crucial role in enabling creative expression through AI-generated outputs. Well-crafted prompts can inspire language models to generate imaginative, engaging, and diverse responses, fostering creativity in storytelling, writing, and other creative applications.
5. **Control and Customization:** Prompts provide a means of controlling and customizing the output generated by language models. By carefully designing prompts, users can influence the style, tone, or specific content of the generated response, allowing for tailored outputs that align with their preferences or specific use cases.
6. **Problem Solving and Assistance:** Prompts can be used to guide language models in problem-solving tasks or providing assistance. By

presenting specific problem statements or queries, prompts enable AI systems to generate relevant solutions, recommendations, or explanations.

7. **Iterative Refinement:** Prompts can be iteratively refined and adjusted to improve the quality or specificity of AI-generated outputs. Through experimentation and fine-tuning of prompts, users can gradually optimize the performance and relevance of the language model's responses.

Overall, prompts serve as powerful tools for influencing and directing AI-generated outputs, allowing users to harness the capabilities of language models effectively and tailor their responses to specific needs, contexts, or creative objectives.

Well-crafted prompts are of significant importance in achieving desired results when working with AI language models. Here are some key reasons why well-crafted prompts are crucial:

1. **Clarity and Specificity:** Well-crafted prompts provide clear and specific instructions or cues to the language model. They communicate the desired outcome or objective in a precise and unambiguous manner. Clear prompts help avoid misinterpretation or ambiguity, ensuring that the language model generates responses that align with the intended purpose.

2. **Guidance and Direction:** Effective prompts guide the language model's attention and focus toward the desired content or context. They provide the necessary cues or constraints to steer the model's understanding and guide the generation process. Well-designed prompts help shape the output by providing explicit instructions or desired characteristics, leading to more relevant and coherent responses.

3. **Contextual Relevance**: Prompts that incorporate relevant contextual information enable language models to generate responses that are contextually grounded. By including appropriate context in the prompts, such as previous dialogue or specific background information, the language model can better understand and produce contextually relevant outputs, enhancing the overall quality and coherence of the generated responses.

4. **Targeting Specific Domains or Styles:** Customized prompts can be tailored to target specific domains, styles, or tones. Well-crafted prompts allow users to influence the language model's behavior and generate outputs that align with specific requirements or preferences. This is particularly useful when seeking domain-specific responses or when aiming to emulate a particular writing style.
5. **Bias Mitigation and Ethical Considerations:** Carefully constructed prompts can help address bias and promote fairness in AI-generated outputs. By being mindful of the prompt wording and instructions, potential biases or controversial content can be minimized. Well-crafted prompts also consider ethical considerations, ensuring responsible AI practices and reducing the risk of generating inappropriate or harmful responses.
6. **Iterative Improvement:** Well-crafted prompts allow for iterative refinement and improvement of AI-generated results. By experimenting with different prompt variations, users can gradually fine-tune the prompts to achieve desired outcomes. This iterative process enables users to optimize and enhance the language model's performance over time.
7. **Creative Exploration:** Prompts can serve as catalysts for creative exploration and imaginative outputs. Well-crafted prompts inspire the language model to generate diverse, engaging, and innovative responses. They encourage the exploration of unique ideas, storytelling techniques, and artistic expression, fostering creativity in AI-generated content.

In summary, well-crafted prompts are crucial for achieving desired results when working with AI language models. They provide clarity, guidance, and context, allowing users to effectively steer the output generation process and tailor the responses to meet specific goals, domains, or creative objectives. Through thoughtful prompt engineering, users can unlock the full potential of AI models and obtain more accurate, relevant, and satisfactory outcomes.

<u>1.2 Prompts as Instructional Signals</u>

Prompts act as instructions to guide language models by providing explicit cues or signals that shape the model's understanding and influence the nature of the generated response. Here's how prompts serve as instructions:

1. **Explicit Instructional Content:** Prompts often contain specific instructions or commands that guide the language model's behavior. These instructions can range from simple directives (e.g., "Translate this sentence into French") to more complex tasks (e.g., "Write a persuasive essay arguing for renewable energy"). By including clear instructions in the prompts, users can guide the language model towards the desired task or outcome.

2. **Formatting and Structure:** The formatting and structure of prompts can provide implicit instructions to language models. For example, by structuring the prompt as a question (e.g., "What are the benefits of exercise?"), the model is prompted to generate a response that answers the question. Similarly, by providing partial information or an incomplete sentence in the prompt, the model can be instructed to complete the sentence or fill in the missing information.

3. **Contextual Cues:** Prompts can include contextual cues that guide the language model's understanding and response generation. By incorporating relevant context, such as previous conversation history or specific background information, prompts provide the necessary context for the model to generate contextually relevant responses. These cues help align the model's understanding with the intended context or topic.

4. **Desired Output Specifications:** Well-crafted prompts can specify the desired characteristics or attributes of the generated output. For instance, prompts can instruct the model to produce responses in a particular style (e.g., "Write a poem in the style of Shakespeare") or with certain constraints (e.g., "Generate a response in 100 words or less"). By providing these specifications, prompts guide the language model towards generating responses that meet the desired criteria.

5. **Domain-Specific Instructions:** Prompts can be tailored to provide domain-specific instructions, guiding the language model to generate outputs specific to a particular field or subject. For instance, prompts can include references to domain-specific terminology, guidelines, or standards, instructing the model to produce outputs that align with that domain's requirements.

6. **Iterative Refinement:** Prompts can be iteratively refined based on the model's previous responses to improve subsequent outputs. By observing and analyzing the model's behavior, users can modify and adjust prompts to better guide the model's understanding and

response generation. This iterative refinement process allows for fine-tuning the instructions and achieving more desired results over time.

In summary, prompts act as instructions to guide language models by providing explicit and implicit cues, context, specifications, and domain-specific guidance. Well-crafted prompts enable users to effectively communicate their desired outcomes, directing the language model's behavior and ensuring the generation of responses that align with the intended instructions and objectives.

Conditioning and priming are important concepts in prompt engineering, playing a crucial role in guiding language models and influencing their response generation. Let's explore these concepts in more detail:

Conditioning: Conditioning refers to the process of influencing a language model's behavior by providing specific cues or conditioning signals in the prompt. These cues condition the model's internal state and guide its understanding and response generation. The conditioning can be achieved through various means, such as explicit instructions, contextual information, or partial input.

For example, a prompt like "Translate the following sentence into Spanish: 'Hello, how are you?'" conditions the language model to understand that the desired task is to perform a translation into Spanish. The conditioning signal of "Translate" guides the model's behaviour and prompts it to generate a response aligned with the translation task.

Conditioning signals can also be used to guide models in generating specific styles or tones. For instance, a prompt like "Write a formal email to a potential client" conditions the model to adopt a formal tone and structure in the generated response.

Priming: Priming involves setting the initial state or context of a language model before generating the response. It influences the model's interpretation of subsequent text and shapes its response accordingly.

Priming is often used to provide contextual cues or reference points that guide the model's understanding and improve the relevance of the generated output.

In prompt engineering, priming can be achieved by incorporating relevant contextual information in the prompt. This information can include preceding dialogue, background knowledge, or specific details related to the desired output. By priming the model with this contextual information, it can generate responses that are more contextually grounded and aligned with the given prompt.

For example, in a conversational context, the prompt "They asked me about my hobbies. I replied, 'I enjoy playing tennis and painting.' Then, they asked..." primes the model by providing preceding dialogue. The model can then generate a response that is consistent with the context established in the prompt.

Priming can also be used to bias the model's response towards a particular topic or perspective. By introducing specific information or framing in the prompt, models can be primed to generate outputs that align with the desired viewpoint.

Overall, conditioning and priming in prompt engineering are powerful techniques to guide language models and shape their response generation. They enable users to provide explicit instructions, establish context, and bias the model's behavior, resulting in more relevant, context-aware, and desired outputs.

1.3 Prompts and Contextual Understanding

Prompts serve as contextual cues for language models, allowing them to generate context-aware responses. By incorporating relevant context in the prompts, language models can better understand the desired context and produce more appropriate and coherent outputs. Here's how prompts act as contextual cues:

1. **Dialogue Context:** In conversational contexts, prompts can include preceding dialogue or conversation history. By referencing previous exchanges, prompts establish the context for the current response generation. Language models can utilize this contextual information to generate responses that align with the ongoing conversation, ensuring continuity and coherence.
2. **Background Information:** Prompts can provide essential background information relevant to the desired response. This information sets the stage and helps the language model understand the specific context in which the response should be generated. By including relevant facts, events, or details, prompts enable the model to consider the broader context and generate contextually appropriate responses.
3. **Specific Task or Scenario:** Prompts can present a specific task, scenario, or situation, providing the necessary context for response generation. By framing the prompt within a particular context, language models can generate outputs that are specifically tailored to that context. For example, a prompt like "You are a tour guide. Provide information about the historical landmarks in the city" sets the context for the model to generate context-aware responses about historical landmarks.
4. **Partial Information or Questions:** Prompts can introduce partial information or ask questions, prompting the language model to generate responses that complete the missing information or answer the question based on the context. This enables the model to understand the context implied by the prompt and generate appropriate responses. For example, a prompt like "The capital of France is..." cues the model to generate a contextually aware response completing the sentence with "Paris."
5. **Contextual Constraints:** Prompts can impose constraints that guide the language model's response generation within a specific context. For instance, a prompt like "Write a news headline about the recent sports event" provides a contextual cue of generating a response that adheres to news headline conventions in the sports domain. This ensures that the model generates responses that are contextually consistent and fit the desired genre or format.

By incorporating contextual cues, prompts enable language models to consider the broader context in which the response is expected. This results in more context-aware and coherent responses, enhancing the relevance and quality of the generated outputs.

Prompts play a crucial role in leveraging pre-trained language models by guiding their understanding and influencing the generated outputs. Here's how prompts influence the utilization of pre-trained language models:

1. **Steering Model Behavior:** Prompts help steer the behavior of pre-trained language models by providing explicit instructions or cues. By carefully crafting prompts, users can guide the model to generate responses that align with their desired outcomes. For example, prompts can instruct the model to summarize a given passage, answer specific questions, or complete a sentence in a particular style. This allows users to leverage the capabilities of pre-trained models to perform specific tasks or generate tailored outputs.
2. **Contextual Understanding:** Pre-trained language models are trained on vast amounts of text data, enabling them to capture patterns, context, and linguistic knowledge. However, prompts provide additional contextual cues that help the model better understand the specific context in which the response is expected. By incorporating relevant context in prompts, such as preceding dialogue or background information, the model can generate contextually grounded responses. This enhances the model's ability to produce coherent and relevant outputs.
3. **Fine-tuning and Transfer Learning:** Prompts serve as input during the fine-tuning process, which involves further training a pre-trained language model on a specific task or domain. By providing task-specific prompts during fine-tuning, the model learns to generate responses tailored to that task. Prompts act as conditioning signals, allowing the model to adapt its behavior and leverage its pre-existing knowledge while refining its performance for the specific task.
4. **Domain Adaptation:** Prompts can be used to leverage pre-trained language models in specific domains. By designing prompts that include domain-specific terminology or context, users can prompt the model to generate outputs that are specific to that domain. This allows the model to leverage its general language understanding while producing domain-specific responses.

5. **Bias Mitigation:** Prompts can be crafted with careful consideration to mitigate bias in the generated outputs of pre-trained language models. By incorporating bias-aware prompts, users can influence the model to generate more fair and unbiased responses. This can help address concerns related to biased or discriminatory language generation.

6. **Iterative Improvement:** Prompts can be iteratively refined to improve the performance of pre-trained language models. By analyzing the generated outputs and modifying the prompts based on the observed behavior, users can iteratively optimize the model's responses. This iterative refinement process allows for better alignment with the desired outputs and continual improvement of the model's performance.

In summary, prompts play a vital role in leveraging pre-trained language models by guiding their behavior, providing contextual understanding, enabling fine-tuning, supporting domain adaptation, mitigating bias, and facilitating iterative improvement. Well-crafted prompts are instrumental in harnessing the capabilities of pre-trained language models and maximizing their usefulness in various applications.

1.4 Prompts and Creative Expression

Prompts have the potential to unleash creativity and generate diverse, imaginative outputs from language models. Here's how prompts can be leveraged to achieve this:

1. **Open-Ended Prompts:** Using open-ended prompts encourages language models to explore different possibilities and generate creative responses. By providing prompts that do not impose strict constraints or specific instructions, the models have the freedom to generate imaginative outputs. Open-ended prompts can include cues like "Imagine a world where..." or "Describe a scenario where..."

2. **Storytelling Prompts:** Prompts that involve storytelling elements can inspire language models to generate narrative-driven responses. By setting the stage, introducing characters, or providing a scenario, prompts can encourage the models to

generate imaginative stories or narratives. For example, prompts like "Once upon a time..." or "In a distant galaxy..."

3. **Alternate Perspectives:** Prompts that encourage language models to adopt alternate perspectives or personas can lead to imaginative outputs. By instructing the model to generate responses as if they were a historical figure, a fictional character, or an alien species, the models can generate unique and imaginative perspectives.

4. **Creative Constraints:** While prompts can be open-ended, introducing creative constraints can also stimulate imaginative outputs. For example, prompts that restrict responses to a certain word count, require the use of specific words or phrases, or ask for responses in a particular format (such as poetry or dialogue) can inspire language models to generate imaginative responses within those constraints.

5. **Conceptual Prompts:** Prompts that involve abstract concepts or thought-provoking ideas can spark creativity in language models. These prompts can explore philosophical questions, abstract concepts, or hypothetical scenarios. By engaging with these prompts, language models can generate imaginative responses that delve into deeper and more unconventional territories.

6. **Visual or Sensory Prompts:** Pairing prompts with visual or sensory stimuli can stimulate the imagination of language models. By providing images, sounds, or other sensory cues alongside prompts, models can generate responses that evoke vivid and imaginative descriptions or experiences.

7. **Iterative Prompt Refinement**: Experimenting with different variations of prompts and iteratively refining them can lead to increasingly imaginative outputs. By observing the generated responses, users can modify and adjust prompts to elicit more imaginative and creative results. This iterative process allows for the exploration of different prompt formulations and encourages language models to generate diverse and imaginative outputs.

In summary, prompts have the potential to unleash the creativity and generate diverse, imaginative outputs from language models. Open-ended prompts, storytelling elements, alternate perspectives, creative constraints, conceptual prompts, sensory cues, and iterative refinement all contribute to fostering creativity in the generated responses. By exploring different prompt strategies and encouraging imaginative thinking, users can unlock the full creative potential of language models.

Certainly! Here are some examples of prompt-driven creative writing, storytelling, and art:

1. Creative Writing:
- Prompt: "Write a short story that begins with the sentence, 'The door creaked open, revealing a hidden world.'"
- Response: A writer could create a story that explores an enchanted realm discovered behind a mysterious door, delving into magical creatures, unexpected adventures, and the protagonist's personal transformation.

2. Storytelling:
- Prompt: "Tell a tale about a mischievous cat who embarks on a quest to find the lost treasure of an ancient civilization."
- Response: Through storytelling, one could craft a whimsical narrative following the adventures of the mischievous cat, encountering various obstacles, and uncovering clues leading to the hidden treasure.

3. Art:
- Prompt: "Create an illustration depicting a futuristic cityscape with floating islands and advanced transportation systems."
- Response: Artists can bring this prompt to life through their artistic interpretation, utilizing colors, shapes, and perspective to portray a visually stunning and imaginative futuristic cityscape.

4. Poetry:
- Prompt: "Compose a poem exploring the beauty and tranquility of a starlit night."
- Response: Poets can craft verses that evoke the ethereal atmosphere of a starlit night, capturing its enchanting qualities, the play of light, and the emotions it inspires.

5. Music:
- Prompt: "Compose a musical piece inspired by the theme of rebirth and renewal."
- Response: Composers can create a musical composition that embodies the essence of rebirth and renewal, utilizing

various instruments, melodies, and harmonies to convey the transformative nature of the theme.

6. Photography:
- Prompt: "Capture a photograph that showcases the juxtaposition of nature and urbanization."
- Response: Photographers can explore their surroundings to find scenes that juxtapose natural elements with urban landscapes, capturing the contrast and harmony between the two.

Remember, these examples are just starting points, and the actual creative outputs will vary based on individual interpretation and artistic style. Prompts provide a spark of inspiration, allowing individuals to unleash their creativity and bring forth their unique artistic expressions in writing, storytelling, and various forms of art.

1.5 The Science behind Prompts: How Language Models Respond

Understanding how language models interpret and generate responses to prompts involves considering the underlying mechanisms and processes. Here's an overview of how language models approach prompt interpretation and response generation:

1. **Pre-trained Knowledge:** Language models, particularly those based on transformer architectures like GPT-3, are trained on massive amounts of text data. This pre-training phase allows models to capture linguistic patterns, grammar, context, and semantic relationships. The acquired knowledge forms the foundation for subsequent prompt interpretation and response generation.
2. **Tokenization:** Language models tokenize input text, breaking it down into smaller units called tokens. Tokens can represent individual words, subwords, or characters. This tokenization process allows the model to handle text inputs efficiently.
3. **Encoding and Contextual Representation:** Once tokenized, the language model applies encoding techniques, such as transformer layers, to create contextual representations of the input tokens. The model captures the relationships between tokens, taking into account the surrounding context and dependencies. This

contextual representation helps the model understand the meaning and intent behind the prompt.

4. **Conditioning and Priming:** Language models use prompts as conditioning signals or priming cues to guide their response generation. The prompts influence the internal state of the model, shaping its understanding and subsequent output. Conditioning can be achieved through explicit instructions, contextual cues, or partial input.

5. **Generation Process:** Language models generate responses through a decoding process. Given the prompt and its contextual representation, the model generates tokens one by one, expanding the output step by step. The model utilizes probabilistic techniques to determine the most likely token to generate at each decoding step, considering both the learned language patterns and the influence of the prompt.

6. **Beam Search and Sampling:** Language models often employ beam search or sampling methods during response generation. Beam search explores multiple possible token sequences, maintaining a list of the most promising candidates based on their likelihood. Sampling, on the other hand, randomly selects tokens according to their probability distribution, allowing for more varied and diverse outputs.

7. **Fine-tuning and Transfer Learning:** Language models can undergo fine-tuning on specific tasks or domains using prompts. During this phase, models are trained on task-specific data with associated prompts, adapting their behavior to perform well on the targeted task. Fine-tuning further refines the model's ability to interpret prompts and generate appropriate responses.

8. **Bias and Ethical Considerations:** It's important to note that language models can exhibit biases present in the training data, and prompts can inadvertently reinforce or amplify those biases. Ethical considerations should be taken into account when crafting prompts and evaluating the generated responses to mitigate potential biases and ensure fairness.

Overall, language models interpret prompts by leveraging their pre-trained knowledge, encoding contextual representations, conditioning on prompts, and utilizing probabilistic generation processes. The response generation involves decoding and beam search/sampling techniques. Through fine-tuning and ethical considerations, prompt interpretation and response generation can be guided and refined to align with desired outcomes and societal considerations.

The role of model architecture, training data, and fine-tuning in prompt processing is crucial for the effective interpretation and generation of responses. Here's an analysis of each component:

1. Model Architecture:

- The model architecture, such as the transformer architecture used in models like GPT-3, plays a significant role in prompt processing. The architecture determines how the model processes and represents input text, capturing contextual dependencies and relationships between tokens.
- The design of the architecture influences the model's ability to understand prompts, encode contextual information, and generate coherent and relevant responses. More advanced architectures often exhibit improved language understanding and generation capabilities, which can enhance prompt processing.

2. Training Data:

- The training data used to pre-train language models has a profound impact on prompt processing. Large-scale datasets consisting of diverse, high-quality text are typically employed.
- The quality, diversity, and representativeness of the training data influence the model's ability to understand prompts from various domains, contexts, and languages. Models trained on diverse datasets can exhibit better generalization and comprehension of prompts during response generation.

3. Fine-tuning:

- Fine-tuning involves training the pre-trained language models on specific tasks or domains using task-specific datasets and associated prompts. Fine-tuning helps adapt the models to specific prompt processing requirements.
- Fine-tuning enables models to specialize in certain tasks or domains, allowing them to understand and generate responses aligned with the desired outcomes. The fine-

tuning process refines the model's prompt interpretation and response generation capabilities based on task-specific prompts and target datasets.

4. Prompt Engineering:

- Prompt engineering refers to the practice of carefully designing prompts to elicit desired behaviors and outputs from language models. It involves considering the choice of words, instructions, context, constraints, and framing to guide the model's interpretation and response generation.
- Well-crafted prompts can significantly influence prompt processing. They can provide explicit instructions, introduce context, set constraints, or prime the model for specific responses. Prompt engineering plays a vital role in leveraging the capabilities of the model architecture, training data, and fine-tuning to achieve the desired prompt processing outcomes.

In summary, model architecture, training data, fine-tuning, and prompt engineering collectively contribute to effective prompt processing. The architecture influences the model's ability to understand and generate responses, while the quality and diversity of training data impact the model's language comprehension. Fine-tuning refines the model's behavior for specific prompt processing tasks, and prompt engineering guides the interpretation and generation of responses. Considering and optimizing these components can enhance the overall prompt processing capabilities of language models.

1.6 The Art of Prompt Design

Crafting effective prompts is crucial for guiding language models and eliciting desired responses. Here are key considerations to keep in mind when crafting prompts:

1. **Clarity and Specificity:** Prompts should be clear, specific, and unambiguous. They should provide precise instructions or cues to guide the language model's behavior. Avoid vague or ambiguous language that can lead to misinterpretation or undesired outputs.

2. **Contextual Information:** Incorporate relevant contextual information in prompts to provide necessary background or context for the desired response. This helps the language model generate context-aware and coherent outputs. Consider including relevant details, preceding dialogue, or specific situational cues to set the appropriate context.

3. **Desired Outcome:** Clearly define the desired outcome or goal for the response. Whether it's answering a specific question, providing a summary, or generating creative content, articulate the intended result to guide the model's behavior.

4. **Length and Format:** Consider the desired length and format of the response. If you require a concise answer, specify word limits or ask for a brief summary. If you want a creative narrative, encourage storytelling elements. Tailor the prompt's length and format to align with the desired response structure.

5. **Task-specific Instructions**: If the prompt is related to a specific task or domain, provide task-specific instructions. Use domain-specific terminology, provide guidelines, or include examples that help the language model produce more accurate and relevant outputs.

6. **Conditioning Signals:** Utilize conditioning signals within the prompt to guide the model's behavior. Explicitly instruct the model to perform certain actions, use specific words or phrases, or adopt a particular style. Conditioning signals can effectively steer the model's interpretation and response generation.

7. **Ethical Considerations:** Consider ethical implications when crafting prompts. Avoid prompts that may lead to biased, offensive, or harmful content generation. Be mindful of potential biases present in the training data and aim for fairness, inclusivity, and responsible AI use.

8. **Iterative Refinement:** Refine and iterate on prompts based on the observed model behavior. Analyze generated responses, identify areas for improvement, and modify prompts accordingly. Iterative refinement allows for fine-tuning prompts to achieve desired outcomes and enhance the model's performance.

9. **Experimentation and Evaluation:** Experiment with different prompt formulations and evaluate the generated responses. Test various prompt variations, conditioning strategies, or contextual cues to explore their impact on the model's behavior. Evaluation helps identify effective prompt designs and iterate towards more successful prompts.

By considering these key considerations, you can craft prompts that effectively guide language models, elicit desired responses, and align with the intended outcomes of your interactions with the model.

When crafting prompts, exploring different structures, formats, and styles can yield diverse and interesting results. Here are some examples to inspire your exploration:

1. Open-Ended Prompts:
- Begin with "Imagine..." or "What if..." to encourage creative and imaginative responses.
- Start with a thought-provoking question that sparks curiosity and invites exploration.

2. Instructional Prompts:
- Use clear instructions like "Describe," "Explain," or "Compare and contrast" to guide the model's response.
- Provide step-by-step prompts that lead the model through a process or problem-solving task.

3. Storytelling Prompts:
- Start with a captivating opening line or a narrative hook that sets the stage for a story.
- Provide prompts that introduce a conflict, challenge, or unexpected event to inspire storytelling.

4. Dialogue Prompts:
- Frame the prompt as a conversation between two or more characters, guiding the model to generate dialogue-based responses.
- Incorporate specific dialogue cues or character attributes to shape the conversation.

5. Descriptive Prompts:
- Encourage the model to provide detailed descriptions of people, places, objects, or events.
- Include sensory cues like sight, sound, smell, taste, and touch to stimulate vivid and immersive descriptions.

6. Opinion or Argument Prompts:
- Ask the model to express its opinion on a particular topic or argue for or against a specific position.
- Frame the prompt in a way that invites the model to provide reasoning, evidence, or counterarguments.

7. Visual Prompts:

- Combine written prompts with visual stimuli like images, artwork, or photographs.
- Ask the model to describe or interpret the visual elements, evoke emotions, or create a story inspired by the image.

8. Constraint-based Prompts:
- Set specific constraints or limitations on the response, such as word count, rhyme scheme, or adherence to a specific style or format (e.g., haiku, sonnet, or limerick).
- Challenge the model to think creatively within the given constraints.

9. Persuasive Prompts:
- Prompt the model to craft persuasive arguments or speeches to convince an audience.
- Include cues for appealing to emotions, presenting evidence, or addressing counterarguments.

10. Collaborative Prompts:
- Frame the prompt as a collaborative task between the model and the user.
- Encourage back-and-forth interactions, alternate responses, or building upon each other's contributions.

Remember, experimentation is key when exploring different prompt structures, formats, and styles. By varying these aspects, you can uncover new creative possibilities, generate diverse outputs, and engage in unique interactions with the language model.

1.7 Real-World Applications of Prompt Engineering

Prompt engineering plays a crucial role in various fields, including chatbots, content generation, and virtual assistants. Here's how prompt engineering is utilized in these domains:

1. Chatbots:
- Chatbots rely heavily on prompt engineering to generate relevant and contextually appropriate responses in conversational interactions.

- Prompt engineering helps guide chatbots to understand user queries, intents, and preferences, enabling them to provide accurate and helpful responses.
- Crafting prompts that elicit specific information or actions from users allows chatbots to collect necessary data and guide users through various processes.

2. Content Generation:

- In content generation, prompt engineering is utilized to instruct language models to produce specific types of content, such as articles, blog posts, or product descriptions.
- Prompts can provide guidelines on desired tone, style, structure, or target audience, ensuring that the generated content aligns with specific requirements.
- By carefully designing prompts, content generators can steer language models to produce coherent, engaging, and on-topic content tailored to their needs.

3. Virtual Assistants:

- Virtual assistants like Siri, Alexa, or Google Assistant rely on prompt engineering to understand user commands and provide relevant information or perform specific tasks.
- Prompt engineering ensures that virtual assistants accurately interpret user prompts and generate appropriate responses or take appropriate actions.
- Crafting prompts that incorporate context, user preferences, and specific instructions enables virtual assistants to deliver personalized and useful assistance.

4. Language Translation:

- Prompt engineering is valuable in language translation tasks, where language models are conditioned with prompts in the source language to generate translations in the target language.
- By carefully framing prompts, translators can guide language models to produce accurate translations, considering context, idiomatic expressions, or desired levels of formality.

5. Customer Support:

- Prompt engineering is instrumental in customer support systems, where language models assist in answering customer queries or resolving issues.

- Prompts can be crafted to gather necessary information from customers, such as order details or specific concerns, to provide more efficient and tailored support.
- By engineering prompts that anticipate common customer inquiries, support systems can generate pre-defined responses or suggest relevant resources, enhancing the customer service experience.

In these fields, prompt engineering serves as a vital technique to guide language models, ensuring accurate interpretation of user inputs, generation of contextually appropriate responses, and alignment with specific objectives or requirements. By carefully designing prompts, professionals in these domains can leverage the power of language models to enhance user experiences, automate processes, and generate high-quality content.

Certainly! Here are a few case studies showcasing successful implementations of prompt engineering:

1. OpenAI's "InstructGPT" for Code Generation:

- OpenAI used prompt engineering to train their language model, InstructGPT, specifically for code generation tasks.
- By providing prompts that resemble natural language descriptions of code requirements, developers can instruct InstructGPT to generate Python code that matches the desired functionality.
- Prompt engineering helped align the model's responses with specific programming tasks, enabling developers to leverage InstructGPT effectively for code generation.

2. Google's "Smart Compose" in Gmail:

- Google implemented prompt engineering in their "Smart Compose" feature in Gmail, which suggests completions for email compositions.
- By analyzing the user's partially written sentence, Smart Compose generates prompts that help users complete their emails quickly.
- Prompt engineering plays a crucial role in understanding the user's context and intent, enabling Smart Compose to provide relevant suggestions and enhance productivity.

3. Storytelling with AI: AI Dungeon:

- AI Dungeon is an interactive storytelling game powered by prompt engineering and language models.
- Users provide story prompts and interact with the AI-generated narrative. The prompts guide the AI model to create context-aware and engaging storylines.
- By carefully crafting prompts that set the scene, introduce characters, or pose challenges, AI Dungeon creates dynamic and immersive storytelling experiences for users.

4. Chatbot Assistants for Customer Service:

- Many businesses employ prompt engineering in their chatbot assistants to enhance customer service interactions.
- By crafting prompts that anticipate common customer queries and guide the chatbot's responses, businesses can ensure accurate and helpful information is provided to customers.
- Prompt engineering helps streamline customer support processes, provide quick resolutions, and enhance the overall customer experience.

These case studies demonstrate how prompt engineering can be successfully implemented across various domains, such as code generation, email composition, storytelling, and customer service. By leveraging well-crafted prompts, organizations and developers can guide language models to generate contextually relevant and valuable outputs, leading to improved productivity, user experiences, and automation capabilities.

1.8 Ethical Considerations in Prompt Engineering

Addressing ethical concerns related to prompts is essential to ensure fairness, mitigate biases, and uphold responsible AI practices. Here are some key considerations to address these concerns:

1. Bias Detection and Mitigation:

- Evaluate prompts for potential biases and address them proactively. Review prompts to ensure they do not reinforce stereotypes, discrimination, or marginalization.
- Implement bias detection techniques to identify and address any biases in the generated responses. Regularly

monitor and update prompts to align with ethical
guidelines and societal standards.

2. Diverse and Inclusive Prompts:

- Craft prompts that encourage diverse perspectives, inclusivity, and respectful language. Consider the potential impact of prompts on different cultural backgrounds, genders, or communities.
- Incorporate diverse examples, contexts, and scenarios to ensure that the language model is exposed to a wide range of experiences and viewpoints.

3. User Feedback and Iterative Improvement:

- Establish mechanisms to collect user feedback on the prompts and generated responses. Actively engage with users to understand their experiences and address any concerns or biases that may arise.
- Iterate and refine prompts based on user feedback, continuously improving the model's performance and reducing biases over time.

4. Transparent Prompt Design:

- Be transparent about the purpose and intent behind prompts. Clearly communicate the expected outcomes and limitations to users interacting with the language model.
- Provide information on how prompts are crafted, evaluated, and updated. Foster trust by promoting transparency and accountability in the prompt engineering process.

5. User Consent and Control:

- Respect user autonomy by allowing them to choose the level of interaction and control over prompts. Provide options to customize or modify prompts to align with individual preferences and values.
- Obtain informed consent from users regarding the use of their data and the prompts employed in generating responses. Clearly communicate the data usage policies and privacy practices.

6. Multistakeholder Collaboration:

- Foster collaboration between researchers, developers, ethicists, and domain experts to collectively address ethical concerns related to prompts.

- Engage in interdisciplinary discussions and incorporate diverse perspectives to ensure a holistic approach towards ethical prompt engineering.

7. Regular Auditing and External Reviews:
- Conduct regular audits and external reviews of prompt engineering practices. Seek independent assessments to evaluate the ethical implications, biases, and fairness of prompts and the generated responses.
- Embrace external scrutiny to identify potential shortcomings, biases, or unintended consequences, and take corrective actions accordingly.

By incorporating these ethical considerations into prompt engineering processes, organizations and researchers can strive for fair, inclusive, and responsible AI practices. Continual evaluation, improvement, transparency, and user-centric approaches are crucial for addressing biases, ensuring fairness, and promoting the responsible use of AI technologies.

Ensuring ethical considerations in prompt design and implementation is vital to promote responsible and unbiased AI practices. Here are some key steps to incorporate ethical considerations into prompt design and implementation:

1. Understand Ethical Implications:
- Familiarize yourself with ethical principles, guidelines, and frameworks relevant to AI and language models.
- Gain awareness of potential biases, fairness concerns, and societal impact associated with prompt design and implementation.

2. Diversity and Inclusion:
- Strive for diversity and inclusivity in prompt design by incorporating a wide range of perspectives, cultures, and backgrounds.
- Avoid prompts that perpetuate stereotypes, discrimination, or marginalization of any individuals or groups.

3. Bias Detection and Mitigation:
- Conduct thorough audits and assessments to identify potential biases in prompts and the resulting generated responses.

- Implement techniques such as bias-correction algorithms or debiasing methods to mitigate any identified biases.

4. User-Centric Approach:

- Place the user's interests, values, and privacy at the forefront of prompt design and implementation.
- Allow users to have control over the prompts presented to the language model and offer options to customize or modify prompts based on their preferences.

5. Informed Consent and Transparency:

- Clearly communicate to users how prompts are used, the limitations of the language model, and the data privacy practices associated with prompt-driven interactions.
- Obtain informed consent from users regarding the use of their data and the prompts utilized in generating responses.

6. Continuous Monitoring and Iterative Improvement:

- Regularly monitor the performance and ethical implications of prompts and generated responses.
- Gather user feedback, conduct audits, and engage in iterative improvement processes to address any ethical concerns that arise.

7. Collaborative and Multidisciplinary Approach:

- Foster collaboration between researchers, ethicists, domain experts, and stakeholders to collectively address ethical considerations in prompt design and implementation.
- Encourage diverse perspectives and interdisciplinary discussions to ensure a holistic approach.

8. External Auditing and Reviews:

- Seek external audits and reviews from independent experts or organizations to assess the ethical implications, biases, and fairness of prompts and the resulting outputs.
- Embrace external scrutiny to identify potential shortcomings, biases, or unintended consequences and take corrective actions accordingly.

By integrating these ethical considerations into prompt design and implementation, organizations and researchers can promote responsible AI practices, uphold fairness, and mitigate biases. Continuous evaluation, user feedback, transparency, and collaboration are key to ensuring ethical prompt design and implementation in AI systems.

<u>1.9 The Future of Prompt Engineering</u>

Prompt engineering is an evolving field with the potential for several advancements and trends. Here are a few potential areas of advancement and emerging trends in prompt engineering:

1. Contextual and Dynamic Prompting:
- Advancements in prompt engineering may focus on enhancing the contextual understanding of prompts. Language models could become more adept at interpreting nuanced prompts and capturing the evolving context within a conversation.
- Dynamic prompting techniques may be developed, where prompts adapt and evolve based on the ongoing dialogue or user inputs, enabling more interactive and context-aware interactions.

2. Personalized Prompts:
- The future of prompt engineering may involve personalized prompts tailored to individual users. AI systems could learn from user preferences, historical interactions, and contextual cues to generate prompts that align with specific user needs and preferences.
- Personalized prompts have the potential to enhance user experiences, improve the relevance of generated responses, and foster a sense of customization.

3. Bias-Aware Prompting:
- Addressing bias in prompt engineering will likely continue to be a prominent focus. Advancements may involve the development of bias-detection algorithms that can analyze prompts for potential biases before generating responses.
- Researchers and practitioners may explore techniques to actively mitigate biases within prompts, ensuring more fair and unbiased language model outputs.

4. Prompt Exploration and Reinforcement Learning:

- Future advancements may involve techniques that allow language models to explore and experiment with prompts to improve their performance iteratively.
- Reinforcement learning approaches could be employed to optimize prompts based on user feedback, reward signals, or specific objectives, leading to prompt engineering techniques that are more efficient and effective.

5. Multimodal Prompts:

- Prompt engineering may expand to incorporate multimodal inputs, such as combining text with images, audio, or video prompts. Language models could be trained to interpret and generate responses based on diverse multimodal prompts, enabling more interactive and immersive interactions.

6. Collaborative Prompt Design:

- The field of prompt engineering may see an increase in collaborative prompt design approaches. Developers, domain experts, and end-users may work together to co-create prompts that align with specific use cases, ensuring a more diverse range of perspectives and expertise.

7. Ethical Prompt Engineering Frameworks:

- As the ethical considerations in AI continue to gain attention, the development of frameworks and guidelines specifically for prompt engineering may emerge.
- These frameworks would provide best practices and standards to address ethical concerns, fairness, transparency, and user-centric design in prompt engineering.

These potential advancements and emerging trends in prompt engineering reflect the ongoing efforts to enhance the capabilities, fairness, and user experiences of AI systems. As the field progresses, prompt engineering techniques will likely evolve to meet the needs of various applications and address the ethical considerations associated with AI technologies.

The role of prompts in the field of natural language processing (NLP) is expected to evolve in several ways. Here are some speculations on the evolving role of prompts:

1. Guiding Language Models:
- Prompts will continue to play a crucial role in guiding language models, helping them understand user inputs, generate contextually relevant responses, and align with specific tasks or objectives.
- As language models become more sophisticated, prompts will provide increasingly nuanced instructions, allowing users to elicit specific types of information or responses from the models.

2. Contextual Understanding:
- Prompts will evolve to provide better context for language models. They will help capture the conversational history, user intent, and contextual cues, enabling models to generate more accurate and coherent responses.
- Context-aware prompts will enhance the overall quality of interactions, making language models more capable of maintaining coherent dialogue or delivering personalized responses.

3. Personalized and Adaptive Prompts:
- The role of prompts may shift towards personalization, where prompts are customized to individual users based on their preferences, history, or specific requirements.
- Adaptive prompts may dynamically change based on user feedback, real-time interactions, or evolving conversation context, leading to more tailored and user-centric language model outputs.

4. Collaborative Prompt Design:
- Prompt design may become a collaborative process involving input from developers, domain experts, and end-users. This collaborative approach ensures that prompts capture diverse perspectives, specific domain knowledge, and address user needs effectively.

5. Bias Mitigation and Fairness:
- The role of prompts will continue to expand in mitigating biases and promoting fairness. Efforts will be made to design prompts that actively counteract biases and promote unbiased language model responses.
- Researchers and practitioners will focus on developing techniques to detect and address biases within prompts to ensure more equitable and fair language model outputs.

6. Exploratory and Interactive Prompts:

- Prompts may become more exploratory and interactive, allowing users to experiment with different prompt variations to guide language models' behavior and generate desired outputs.
- Interactive prompts may enable users to provide iterative feedback, iterate on prompt design, or fine-tune language models in real-time, fostering a more interactive and collaborative user experience.

7. Integration with Other Modalities:

- Prompts will likely evolve to encompass multimodal inputs, integrating text with other modalities such as images, audio, or video.
- Language models will be trained to interpret and generate responses based on diverse multimodal prompts, enabling more comprehensive and multimodal natural language processing capabilities.

These speculations highlight the evolving role of prompts in NLP, reflecting the ongoing advancements in AI technologies, user expectations, and ethical considerations. As the field progresses, prompts will continue to serve as a powerful tool for guiding language models, enhancing user experiences, and ensuring responsible and effective deployment of AI systems in various applications.

1.10 Conclusion

Well-crafted prompts hold significant importance in several aspects of prompt engineering. Here is a recap of their significance:

1. **Guiding Language Models:** Well-crafted prompts serve as instructions to guide language models in understanding user inputs and generating contextually relevant responses. They help shape the behavior and output of language models, aligning them with specific tasks or objectives.
2. **Achieving Desired Results:** Effective prompts are instrumental in achieving the desired results from language models. They provide the necessary information, context, and constraints to elicit

responses that meet user expectations and fulfill specific requirements.

3. **Context Awareness:** Well-crafted prompts enable language models to better understand the conversational context, user intent, and relevant cues. They enhance the models' ability to generate coherent and meaningful responses by considering the broader context of the conversation.

4. **Bias Mitigation:** Carefully designed prompts can contribute to mitigating biases in language model outputs. By avoiding biased language, stereotypes, or discriminatory content in prompts, developers can influence more fair and unbiased responses from the models.

5. **Personalization and User Experience:** Customized prompts tailored to individual users can enhance personalization and improve the overall user experience. Well-crafted prompts consider user preferences, historical interactions, and specific needs, leading to more relevant and engaging language model outputs.

6. **Creative Expression and Artistic Outputs: Prompts** play a crucial role in creative writing, storytelling, and artistic applications powered by language models. They provide the initial spark, context, or constraints to inspire imaginative and diverse outputs, enabling users to explore their creative potential.

7. **Ethical Considerations:** Thoughtfully designed prompts address ethical concerns such as bias, fairness, and responsible AI practices. By considering the ethical implications of prompts, developers can ensure that the language model's responses adhere to ethical guidelines and societal standards.

Overall, well-crafted prompts are fundamental in prompt engineering as they direct language models, shape their outputs, enhance context awareness, mitigate biases, enable personalization, foster creativity, and promote ethical and responsible AI practices. They are essential in achieving the desired outcomes and ensuring the effective and ethical use of language models in various applications.

Chapter 2: The Role of Prompts in Natural Language Processing: A Comprehensive Overview

Chapter 2.1: Understanding Prompts and their Significance

- In the context of natural language processing (NLP), prompts refer to textual or multimodal inputs that serve as instructions or cues given to a language model or AI system to guide its understanding and generate appropriate responses. Prompts can take various forms, including single sentences, questions, instructions, or even a combination of text and other modalities such as images, audio, or video.

- Prompts provide the initial context and guidance for the language model, setting the stage for it to process and interpret subsequent user inputs. They help shape the behavior and output of the model by providing specific task-related information, constraints, or examples that direct its understanding and response generation.

- The effectiveness of prompts in NLP depends on their clarity, relevance, and ability to convey the desired information or task requirements to the language model. Well-crafted prompts play a crucial role in eliciting accurate, contextually appropriate, and meaningful responses from the model, enhancing the overall performance and user experience.

- Prompts are widely used in various NLP applications such as chatbots, virtual assistants, content generation, sentiment analysis, text classification, machine translation, and more. They are essential tools in harnessing the capabilities of language models and guiding them to produce desired outputs in a wide range of language processing tasks.

The role of prompts in guiding language models is fundamental to their operation and plays a crucial role in shaping their responses. Here are the key aspects of the role of prompts in guiding language models:

1. **Contextual Understanding:** Prompts provide the initial context for language models to understand the user's input and generate contextually appropriate responses. By setting the context, prompts help language models interpret subsequent user queries or statements in a meaningful way.
2. **Task Specification:** Prompts specify the task or objective for the language model. They define the desired output or behavior expected from the model, guiding it to generate responses that align with the task's requirements. This helps ensure that the language model produces relevant and task-specific responses.
3. **Constraints and Guidelines:** Prompts can include constraints or guidelines that shape the language model's behavior. These constraints may include restrictions on the response length, explicit instructions on the desired tone or style, or limitations on the type of information to be included in the response. Prompts help guide the language model's decision-making process within predefined boundaries.
4. **Examples and Demonstrations:** Prompts can include examples or demonstrations that illustrate the desired output or behavior. By providing concrete examples, prompts guide language models to generate responses that are similar to the provided examples, ensuring consistency and desired output quality.
5. **Bias Mitigation:** Prompts can be designed to mitigate biases in language model responses. They can explicitly instruct the model to avoid biased language, stereotypes, or discriminatory content. By incorporating fair and unbiased prompts, language models can be guided to generate more equitable and inclusive responses.
6. **Creative and Imaginative Outputs:** Prompts can serve as creative stimuli for language models. They can inspire imaginative and diverse responses by providing prompts that encourage exploration, creativity, and the generation of novel ideas. Well-crafted prompts can unlock the creative potential of language models, enabling them to produce engaging and unique outputs.

Overall, prompts play a vital role in guiding language models by providing context, specifying tasks, setting constraints, offering examples, and shaping the models' behavior. They help language models understand user inputs and generate responses that align with the desired objectives, ensuring relevance, coherence, and adherence to specific guidelines or constraints. By effectively utilizing prompts, developers can shape the output of language models to meet the desired requirements and enhance the user experience in various NLP applications.

Prompts are essential in capturing user intent, context, and task requirements in natural language processing (NLP). Here's a closer look at their importance:

1. **User Intent:** Prompts help capture the user's intent by providing explicit instructions or cues that guide the language model in understanding what the user wants to achieve or inquire about. They frame the user's query or statement in a way that conveys their intention effectively, enabling the language model to generate relevant and accurate responses.
2. **Contextual Understanding:** Prompts provide contextual information that aids the language model in understanding the meaning and nuances of the user's input. By setting the context, prompts provide essential details, background information, or references that shape the interpretation of subsequent user queries. This enables the language model to generate contextually appropriate responses that align with the user's needs.
3. **Task Requirements:** Prompts define the specific task or objective at hand. They communicate the requirements, constraints, or guidelines associated with the task, guiding the language model to produce outputs that fulfill those requirements. Prompts help focus the language model's attention and guide its response generation process towards the desired outcome.
4. **Ambiguity Resolution:** Prompts play a crucial role in resolving ambiguity in user inputs. They provide additional information or context that helps disambiguate potentially ambiguous queries or statements. By clarifying the intended meaning, prompts enable the language model to generate responses that are accurate and aligned with the user's intended interpretation.
5. **Adaptability to Different Domains:** Prompts are highly adaptable to different domains or topics of interest. By tailoring prompts to specific domains, prompts capture the domain-specific vocabulary, concepts, or expectations, allowing the language model to generate more domain-relevant responses. This adaptability ensures that the language model can effectively handle a wide range of user queries and tasks.
6. **Personalization:** Prompts can be customized to individual users or user groups, capturing their preferences, history, or specific needs.

Personalized prompts enhance the language model's ability to generate tailored responses that align with the user's unique requirements and enhance the overall user experience.

By effectively capturing user intent, context, and task requirements, prompts empower language models to generate responses that are accurate, relevant, and aligned with the user's needs. They bridge the gap between the user's input and the language model's output, ensuring that the generated responses are contextually appropriate and fulfill the specific objectives of the NLP task at hand.

<u>Chapter 2.2: Types and Formats of Prompts</u>

Prompts can take various forms, each serving a specific purpose in guiding language models. Here are different types of prompts commonly used in natural language processing (NLP):

1. **Single-Sentence Prompts:** Single-sentence prompts are concise and self-contained instructions or queries that provide the necessary context for the language model to generate a response. They are often used in tasks such as text completion, sentiment analysis, or short-text generation.

Example: "Write a creative ending for the following sentence: 'The sun set behind the mountains, casting a warm glow over...'"

2. **Multi-Turn Prompts:** Multi-turn prompts consist of a series of user utterances or instructions that occur in a conversational context. These prompts capture the back-and-forth interaction between a user and a system, enabling the language model to generate responses that consider the dialogue history.

Example: User: "What's the weather like today?" System: "It's sunny with a high of 25 degrees Celsius. Do you need any more information?" User: "Will it rain tomorrow?"

3. **Conditional Prompts:** Conditional prompts provide explicit conditions or constraints for the language model to follow when

generating a response. They guide the language model to produce output that aligns with the specified conditions.

Example: "Generate a paragraph describing the advantages and disadvantages of renewable energy sources."

4. **Instruction-Based Prompts:** Instruction-based prompts provide clear and specific instructions to the language model on how to generate the desired response. They outline the desired structure, content, or format of the output.

Example: "Write a step-by-step guide on how to bake a chocolate cake, including a list of ingredients and instructions for each step."

5. **Question-Based Prompts:** Question-based prompts involve posing a question to the language model, guiding it to generate a response that answers the question or provides relevant information.

Example: "What are the symptoms of COVID-19, and how can it be prevented?"

6. **Image or Text-Conditioned Prompts:** These prompts incorporate additional visual or textual information alongside the prompt text to guide the language model's response generation. They leverage multimodal inputs to enhance the context and guide the model's understanding.

Example: "Describe the image below in one paragraph: [Image of a beach at sunset]"

These are just a few examples of the different types of prompts used in NLP. The choice of prompt type depends on the specific task, desired output, and the information needed to guide the language model effectively. By employing various types of prompts, developers can shape the language model's responses and achieve the desired results in a wide range of NLP applications.

Prompts can be structured in different formats to guide the language model's response generation process. Here's an exploration of various prompt formats commonly used in natural language processing (NLP):

1. **Fill-in-the-Blank Prompts:** These prompts involve providing a partially completed sentence or phrase, leaving a blank for the language model to fill in. This format encourages the model to generate the missing information or complete the given statement.

Example: "The capital city of France is _____."

2. **Question-Based Prompts:** Question-based prompts involve posing a question to the language model, guiding it to generate a response that answers the question or provides relevant information.

Example: "What are the main causes of climate change?"

3. **Instruction-Based Prompts:** Instruction-based prompts provide clear and specific instructions to the language model on how to generate the desired response. They outline the desired structure, content, or format of the output.

Example: "Write a persuasive essay arguing for or against the use of genetically modified organisms (GMOs) in agriculture."

4. **Scenario-Based Prompts:** Scenario-based prompts present a specific situation or scenario to the language model, prompting it to generate a response based on the given context.

Example: "You are a customer service representative. Respond to a customer complaint about a delayed shipment and offer a resolution."

5. **Comparative Prompts:** Comparative prompts involve comparing or contrasting different entities, concepts, or options, guiding the language model to generate responses that highlight the similarities or differences between them.

Example: "Compare and contrast the advantages and disadvantages of online shopping versus traditional in-store shopping."

6. **Image- or Text-Conditioned Prompts:** These prompts incorporate additional visual or textual information alongside the prompt text to guide the language model's response generation. The prompt is based on or accompanied by relevant images, descriptions, or context, enhancing the model's understanding and output generation.

Example: "Describe the emotions and story behind this image: [Image of a person standing alone in a rainy street]."

7. **Constraint-Based Prompts:** Constraint-based prompts provide specific constraints or limitations that the language model must adhere to when generating a response. These constraints can include word limits, style guidelines, or content restrictions.

Example: "Write a haiku poem about nature using only three lines and following the 5-7-5 syllable pattern."

These are just a few examples of prompt formats used in NLP. The choice of prompt format depends on the specific requirements of the task, the desired output, and the information needed to guide the language model effectively. By employing different prompt formats, developers can effectively guide language models and shape their responses in various NLP applications.

Different prompt types and formats have varying degrees of suitability for specific NLP tasks and applications. Here's an analysis of how certain prompt types and formats can be suitable for different scenarios:

1. **Fill-in-the-Blank Prompts:** Fill-in-the-blank prompts are effective for tasks that require language completion or generation of missing information. They can be useful for language modeling, text generation, or sentence completion tasks.

2. **Question-Based Prompts:** Question-based prompts are well-suited for information retrieval tasks, question answering systems, or conversational agents. They explicitly guide the language model to

generate responses that address the specific query or provide relevant information.

3. **Instruction-Based Prompts:** Instruction-based prompts are valuable when clear guidelines and structured responses are desired. They work well for tasks such as text summarization, essay writing, or content generation, where the desired output should adhere to specific instructions or follow a particular format.

4. **Scenario-Based Prompts:** Scenario-based prompts are effective in simulating real-world situations or generating context-specific responses. They are useful for chatbots, virtual assistants, or interactive storytelling, where the language model needs to generate coherent and context-aware responses.

5. **Comparative Prompts:** Comparative prompts are suitable for tasks that involve comparing and contrasting entities, options, or concepts. They work well for opinion mining, sentiment analysis, or product reviews, where the language model needs to analyze and provide insights on different aspects of comparison.

6. **Image- or Text-Conditioned Prompts:** Image- or text-conditioned prompts are valuable for multimodal tasks where both visual and textual information is essential. They are useful for image captioning, visual question answering, or generating textual descriptions based on visual inputs.

7. **Constraint-Based Prompts:** Constraint-based prompts are suitable for tasks that require specific limitations or guidelines to be followed. They work well for tasks like text generation with word limits, poetry writing with specific patterns, or style-controlled text generation.

When selecting the most suitable prompt type or format for a specific NLP task or application, it's important to consider the desired output, the nature of the input data, and the specific requirements of the task. Evaluating the task objectives, expected responses, and any constraints will help in determining the optimal prompt type or format that aligns with the desired outcomes and enhances the overall performance of the language model.

Chapter 2.3: Prompt Engineering Techniques

Designing effective prompts is crucial to guide language models and elicit desired responses. Here are some techniques for designing effective prompts:

1. **Capture Context:** Prompts should provide sufficient context to help the language model understand the desired task or query. Contextual information can include relevant background details, previous dialogue history, or specific references to set the context accurately.

Example: "Given the following conversation between two friends, continue the dialogue from Friend A's perspective: [Previous conversation snippet]."

2. **Specify Constraints:** If specific constraints or requirements need to be met, clearly outline them in the prompts. Constraints can include word limits, required topics or themes, style guidelines, or any other limitations that guide the language model's response generation.

Example: "Write a product review in 100 words or less, highlighting the pros and cons of the item."

3. **Provide Examples:** Including examples in prompts can help illustrate the desired output and provide a reference for the language model. Examples can serve as templates, demonstrating the expected format, structure, or content of the generated response.

Example: "Write a short story in the style of Edgar Allan Poe's 'The Tell-Tale Heart,' focusing on themes of guilt and obsession."

4. **Offer Partial Input:** Instead of providing a complete prompt, consider giving partial input to the language model and task it with completing or expanding the provided text. This technique can be useful for tasks like text completion or sentence expansion.

Example: "Complete the sentence: 'In a world where _____.'"

5. **Incorporate User Intent:** To capture user intent effectively, prompts should align with the user's goals or desired outcomes. Understand the user's perspective and frame the prompt in a way that reflects their intention accurately.

Example: "Ask the language model a question about the recent advancements in artificial intelligence."

6. **Adjust Prompt Style:** Adapt the prompt style to match the expected response. For tasks like storytelling or generating creative outputs, use prompts that stimulate imagination and creativity. For factual information retrieval, adopt prompts that seek concise and specific responses.

Example: "Write a poem about love, using metaphor and vivid imagery."

7. **Iterate and Refine:** Designing effective prompts often involves an iterative process. Experiment with different prompt variations, evaluate the output quality, and refine the prompts based on the model's performance and user feedback. Continuous refinement helps optimize the prompts for better results.

By employing these techniques, prompt designers can create effective prompts that guide language models appropriately, elicit desired responses, and achieve the intended outcomes in various NLP tasks and applications.
Conditioning and priming strategies play a significant role in guiding language models' responses. These strategies involve providing specific information or context to the model to shape its output. Here are some techniques:

1. **Text Conditioning:** By conditioning the language model on specific text inputs, such as keywords, phrases, or sentences, you can guide it to generate responses that align with the provided information. Conditioning can be done at the beginning, middle, or end of the prompt to influence the model's behavior.

Example: "Given the sentence 'Once upon a time,' continue the story..."

2. **Contextual Priming:** Priming involves exposing the language model to relevant context or information that can influence its subsequent responses. By priming the model with a specific context, you can bias its behavior and make it more likely to generate responses that are consistent with the given information.

Example: "In the previous conversation, Person A mentioned their love for sports. Respond to Person A's statement with a sports-related comment."

3. **Explicit Instruction:** Providing explicit instructions or directives in the prompt can guide the language model's response generation. Instructions can include specific formatting requirements, tone, style, or any other guidelines to shape the output.

Example: "Write a persuasive essay arguing for the importance of renewable energy sources. Use factual evidence and a formal tone."

4. **Pretraining and Fine-tuning**: Conditioning can also be achieved through the process of pretraining and fine-tuning the language model on specific datasets. By training the model on relevant data, you can bias its knowledge and responses towards specific domains or tasks.

Example: Pretraining a language model on medical literature to generate context-specific responses for healthcare-related prompts.

5. **Multi-Modal Conditioning:** In addition to text, you can also condition language models on other modalities, such as images, audio, or video. Multi-modal conditioning provides additional context and information for the model to generate more accurate and relevant responses.

Example: Providing an image along with a textual prompt to guide the language model's response generation.

6. **Reinforcement Learning:** Conditioning can be accomplished through reinforcement learning techniques, where the language model receives feedback or rewards based on the quality or

relevance of its generated responses. By reinforcing desired behavior, the model can be guided towards more suitable outputs.

Example: Using a reward mechanism to train a chatbot to provide helpful and informative responses.

These conditioning and priming strategies can help guide language models to generate more contextually appropriate and relevant responses in a wide range of NLP tasks and applications. Careful selection and application of these techniques can significantly improve the control and quality of the model's outputs.

Fine-tuning and customizing prompts are essential techniques for improving the performance of language models and tailoring their outputs to specific needs. Here's how you can fine-tune and customize prompts:

1. **Iterative Refinement:** Fine-tuning prompts involves an iterative process of experimentation and refinement. Start with a basic prompt and evaluate the model's responses. Analyze the strengths and weaknesses of the outputs and iteratively modify the prompt to enhance the desired qualities of the generated responses.

2. **Task-Specific Prompt Engineering:** Tailor the prompt specifically for the task or application at hand. Understand the requirements, constraints, and objectives of the task, and design prompts that provide clear instructions or guidelines to the language model. This can involve incorporating domain-specific terminology, providing context, or specifying the desired output format.

3. **Adapting to User Preferences:** Customizing prompts allows you to align the model's outputs with user preferences. Consider incorporating user feedback, preferences, or specific requirements into the prompts to ensure the generated responses cater to their needs. This can involve adjusting the prompt style, language, or tone to match user expectations.

4. **Language Model Interpretability:** Gain insights into how language models interpret prompts by analyzing their behavior and response patterns. Observe how different prompt variations affect the model's outputs and make informed decisions on refining prompts based on the desired outcomes and limitations.

5. **Prompt Length and Structure:** Experiment with different prompt lengths and structures to find the optimal balance. Longer prompts can provide more context but might introduce noise, while shorter prompts may lack clarity. Explore different prompt structures, such as single-sentence prompts, multi-turn prompts, or conditional prompts, to determine which format yields the desired outputs.
6. **Incorporating Human Demonstrations:** To customize the language model's behavior, consider providing demonstrations or examples of desired responses. Fine-tune the model by conditioning it on high-quality human-generated examples that align with the desired outputs. This approach can help the model learn from human expertise and generate more tailored responses.
7. **Monitoring and Feedback:** Continuously monitor the model's performance and collect user feedback to identify areas for improvement. Solicit feedback from users and domain experts to evaluate the quality, relevance, and effectiveness of the model's outputs. Incorporate this feedback to refine and customize prompts accordingly.

Remember that fine-tuning and customizing prompts require careful evaluation, experimentation, and domain expertise. The process involves an ongoing feedback loop to iteratively improve the model's performance and generate tailored outputs that align with the specific requirements and objectives of the task or application.

Chapter 2.4: Prompting Strategies for Various NLP Tasks

Prompts can be adapted and tailored for different NLP tasks to optimize the performance and generate desired outputs. Here's how prompts can be adapted for some common NLP tasks:

1. Text Classification:
- Single-sentence prompts: Use a single-sentence prompt to classify the input text into predefined categories or labels.
- Instruction-based prompts: Provide explicit instructions for the model to classify the text based on specific criteria or features.

- Question-based prompts: Frame the prompt as a question asking the model to identify the category or label that best fits the input text.

2. Text Generation:

- Instruction-based prompts: Provide specific instructions on what type of text to generate, the format, or style desired.
- Conditional prompts: Include a starting sentence or partial text to guide the model's generation, allowing it to continue from that point.
- Scenario-based prompts: Set the context and specify the scenario in which the generated text should be situated.

3. Sentiment Analysis:

- Question-based prompts: Pose a question about the sentiment of the given text, guiding the model to analyze and express its sentiment.
- Comparative prompts: Compare the sentiment of two texts or aspects, instructing the model to evaluate and express the sentiment difference.
- Instruction-based prompts: Provide clear instructions to the model to determine and generate sentiment analysis results for the given text.

4. Machine Translation:

- Conditional prompts: Provide the source language sentence as the initial part of the prompt and instruct the model to generate the translated version.
- Instruction-based prompts: Specify the desired translation criteria, such as maintaining the original meaning, adapting to a specific style, or considering cultural nuances.
- Multi-turn prompts: Utilize a series of prompts and responses to facilitate an interactive translation process, where the model can generate translated output iteratively.

5. Named Entity Recognition:

- Single-sentence prompts: Provide the input text as a prompt and instruct the model to identify and classify named entities within the text.
- Instruction-based prompts: Specify the desired format or type of named entities to recognize (e.g., persons, organizations, locations).

- Contextual prompts: Include contextual information to guide the model in identifying named entities based on the surrounding text.

6. Question Answering:

- Question-based prompts: Frame the prompt as a question and instruct the model to generate the answer based on the provided context or passage.
- Contextual prompts: Include relevant background information or supporting text to guide the model in generating accurate and contextually appropriate answers.
- Multi-turn prompts: Utilize a series of prompts and responses to facilitate a dialogue-based question-answering process, allowing the model to generate responses in a conversational manner.

Adapting prompts for specific NLP tasks involves considering the nature of the task, the desired outputs, and the specific requirements of the application. By tailoring prompts to the task at hand, language models can be guided to produce more accurate and relevant results. Experimentation, evaluation, and fine-tuning are often necessary to optimize prompt adaptation for different NLP tasks.

Certainly! Here are a few case studies that highlight successful prompt engineering approaches for specific NLP tasks:

1. **GPT-3 for Creative Writing:** OpenAI's GPT-3 language model has been used to generate creative written content. By providing context and conditional prompts, users have been able to harness the model's language generation capabilities for various creative writing tasks, such as poetry, short stories, and even screenplay dialogues. The prompts provide initial guidance, allowing users to explore their creativity while leveraging the model's language proficiency.
2. **Google's BERT for Question Answering:** Google's Bidirectional Encoder Representations from Transformers (BERT) model has achieved notable success in question answering tasks. By using well-crafted prompts that incorporate the input question along with contextual information, BERT can effectively generate accurate

answers. The prompts condition the model to pay attention to relevant parts of the input text and produce informative responses.

3. **Microsoft's ChatGPT for Customer Support:** Microsoft has utilized the ChatGPT model for customer support applications. By customizing prompts to include relevant customer queries, context, and instructions, the model can generate appropriate responses to address customer concerns and inquiries. The prompts are designed to capture the specific intent, context, and requirements of customer support interactions, leading to improved customer satisfaction and response quality.

4. **OpenAI's CLIP for Image Captioning**: OpenAI's CLIP (Contrastive Language-Image Pretraining) model has demonstrated success in generating descriptive captions for images. By fine-tuning the model with prompts that combine visual and textual information, CLIP can generate captions that accurately describe the content of the given images. The prompts leverage the model's cross-modal understanding to produce meaningful and contextually relevant image captions.

These case studies highlight the effectiveness of prompt engineering approaches in various NLP tasks. By carefully designing prompts that capture the task requirements, context, and user intent, language models can be guided to generate outputs that align with specific applications and yield successful results.

When selecting and adapting prompts for an NLP task, it's important to consider the characteristics of the task itself. Here are some considerations to keep in mind:

1. **Task Objective:** Understand the objective of the NLP task. Are you aiming for text classification, text generation, sentiment analysis, machine translation, or any other specific goal? The prompt should align with the task objective and guide the model towards generating the desired output.

2. **Input Data Format:** Consider the format of the input data. Is it a single sentence, a paragraph, a document, or a combination of different modalities (text, image, audio, etc.)? The prompt should be designed to accommodate the input data format and capture the necessary information.

3. **Contextual Information:** Determine if contextual information is crucial for the task. Should the prompt capture the context of

previous interactions, historical data, or surrounding text? Incorporating contextual cues in the prompt can help guide the model to generate more accurate and context-aware responses.

4. **Constraints and Requirements:** Identify any specific constraints or requirements for the task. For instance, if generating text, consider constraints like length limits, required format, or domain-specific vocabulary. Specify such constraints within the prompt to guide the model accordingly.

5. **Training Data Availability:** Assess the availability of training data specific to the task. If task-specific training data is limited, consider using prompts to provide additional guidance or incorporate domain-specific information to compensate for the lack of task-specific data.

6. **User Interaction:** Determine if the NLP task involves user interaction, such as in chatbot systems or question-answering scenarios. In such cases, prompts may need to facilitate conversational interactions and prompt the model to respond appropriately to user inputs.

7. **Evaluation Metrics:** Define the evaluation metrics for the task. What metrics will be used to measure the performance and success of the model? Ensure that the prompts are designed to encourage the model to generate outputs that align with the evaluation metrics.

8. **Ethical Considerations:** Take into account ethical considerations such as bias, fairness, and responsible AI practices. Ensure that prompts are designed to minimize bias and promote fair and unbiased outputs.

By considering these aspects specific to the NLP task, you can select and adapt prompts that effectively guide the language model and enable it to generate outputs that align with the task's objectives and requirements. It's essential to iterate, experiment, and evaluate the prompts' effectiveness to refine and improve their performance over time.

Chapter 2.5: Evaluating and Refining Prompts

Evaluating the effectiveness of prompts is crucial to assess their impact on the performance and outputs of language models. Here are some metrics and methodologies to consider when evaluating prompts:

1. **Task-Specific Metrics:** Define task-specific metrics that directly measure the desired outcomes. For example, in text classification tasks, metrics like accuracy, precision, recall, and F1 score can evaluate the effectiveness of prompts in guiding the model to classify inputs correctly. In text generation tasks, metrics such as perplexity, BLEU score, or human evaluation can assess the quality and relevance of the generated outputs.
2. **Comparative Evaluation:** Perform a comparative evaluation of different prompt variations. Compare the performance and outputs of models using different prompts to identify the most effective prompt design. This can involve conducting A/B tests or using techniques like cross-validation to assess the impact of prompts on the model's performance.
3. **Human Evaluation:** Incorporate human evaluators to assess the quality and relevance of the model's outputs generated with different prompts. This can involve creating evaluation sets or scenarios where human judges rate the outputs based on predefined criteria. Human evaluation can provide valuable insights into the effectiveness of prompts in generating outputs that align with human expectations.
4. **Diversity and Novelty:** Consider metrics that measure the diversity and novelty of the generated outputs. For tasks like text generation or creative writing, metrics like diversity scores (e.g., distinct n-grams) or topic coverage can assess the effectiveness of prompts in producing diverse and imaginative outputs.
5. **Bias and Fairness Evaluation:** Evaluate the prompts and their impact on mitigating bias and ensuring fairness. Assess the model's outputs for biases related to gender, race, or other sensitive attributes. Quantitative measures like bias scores or qualitative analysis can help identify and address biases introduced through prompts.
6. **User Feedback and Satisfaction:** Collect user feedback and satisfaction ratings to evaluate the effectiveness of prompts. Solicit feedback from end-users or domain experts to assess the relevance, clarity, and usefulness of the model's responses generated with different prompts. User feedback can provide valuable insights into the practical effectiveness and user experience of prompts.
7. **Real-World Application Performance**: Assess the performance of prompts in real-world applications or deployments. Measure the

impact of prompts on key performance indicators specific to the application, such as customer satisfaction, response time, or conversion rates. Real-world evaluation can provide a comprehensive assessment of the prompts' effectiveness in practical settings.

When evaluating prompts, it's important to use a combination of quantitative and qualitative measures to capture different aspects of their effectiveness. Consider the specific objectives and requirements of the task, and select appropriate evaluation methodologies and metrics that align with those objectives. Additionally, it's essential to iterate and refine prompts based on evaluation results to continually improve their effectiveness.

Refining and optimizing prompts based on user feedback and evaluation results is an iterative process that allows you to improve the effectiveness and performance of prompts over time. Here are some steps to follow in this iterative process:

1. **Gather User Feedback:** Collect feedback from users who have interacted with the language model using different prompts. This feedback can be obtained through surveys, user interviews, or user testing sessions. Understand their experiences, preferences, and any issues they encountered while interacting with the model.
2. **Analyze Evaluation Results:** Evaluate the performance of prompts using metrics, methodologies, and evaluation techniques discussed earlier. Analyze the results to identify areas where prompts can be refined or optimized to improve the model's outputs. Pay attention to specific strengths and weaknesses observed during the evaluation process.
3. **Identify Pain Points and Improvement Opportunities:** Based on the user feedback and evaluation results, identify pain points, shortcomings, or areas where the prompts can be enhanced. This could include addressing biases, improving clarity, providing more relevant context, or accommodating specific user requirements. Prioritize the identified improvement opportunities based on their impact and feasibility.
4. **Iterative Prompt Refinement:** Make incremental changes to the prompts based on the feedback and evaluation results. Experiment

with different prompt variations, such as adjusting the wording, changing the format, incorporating additional instructions, or modifying the context provided. Iterate on the prompts, one modification at a time, to observe the impact on the model's outputs.

5. **Test and Evaluate:** Test the refined prompts by deploying them with the language model. Evaluate the model's performance using appropriate metrics and methodologies. Compare the results with the previous evaluation to assess the effectiveness of the prompt refinements. Seek feedback from users to gather insights on the impact of the changes.

6. **Incorporate User Feedback:** Continue to gather user feedback during the testing phase with the refined prompts. Consider user suggestions and preferences, and use them as inputs for further refinement. Engage in conversations with users to understand their needs better and tailor the prompts accordingly.

7. **Iterate and Improve:** Iterate through steps 3 to 6, continuously refining and optimizing the prompts based on user feedback and evaluation results. This iterative process allows you to gradually enhance the prompts, addressing identified issues and improving the model's performance and user experience.

8. **Monitor Long-term Performance:** Once you have refined and optimized the prompts, monitor the long-term performance of the language model. Track relevant metrics, user feedback, and other performance indicators over time. Make adjustments to the prompts if necessary based on ongoing monitoring and feedback.

By following this iterative process, you can refine and optimize prompts based on user feedback and evaluation results, gradually improving their effectiveness and ensuring that the language model generates more accurate, context-aware, and tailored outputs. Continuous monitoring and iteration are key to maintaining high-quality prompts that align with user needs and application requirements.

Continuous monitoring and improvement in prompt design and implementation are crucial for several reasons:

1. **Adaptation to Changing Needs:** User needs, preferences, and requirements can evolve over time. Continuous monitoring allows you to stay attuned to these changes and adapt the prompts accordingly. By monitoring user feedback and evaluating the

prompt performance, you can identify areas for improvement and make necessary adjustments to meet the changing needs of users.

2. **Addressing Bias and Fairness:** Bias and fairness issues can arise in language models' outputs, and prompts play a role in mitigating these concerns. Continuous monitoring helps identify any biases or unfairness introduced through prompts and prompts' responses. By regularly evaluating and refining prompts, you can address and rectify any biases that may emerge, promoting more equitable and unbiased language model outputs.

3. **Enhancing Effectiveness and Performance:** Monitoring the effectiveness of prompts allows you to identify areas of improvement and optimize their design. By evaluating the prompts' impact on the model's outputs and performance metrics, you can refine and iterate on the prompts to enhance their effectiveness. This leads to improved performance, better alignment with user expectations, and more accurate and relevant responses from the language model.

4. **Iterative Refinement**: Prompt design and implementation benefit from an iterative approach. Continuous monitoring allows you to gather feedback, evaluate results, and refine prompts in an ongoing cycle. Iterative refinement helps to fine-tune prompts over time, considering user feedback, evaluation outcomes, and the evolving needs of the application or task. This iterative process leads to prompt designs that become increasingly effective and tailored to specific requirements.

5. **Ensuring User Satisfaction:** Prompt design and implementation directly impact the user experience. Continuous monitoring helps gauge user satisfaction with the language model's outputs generated using prompts. By actively seeking and incorporating user feedback, you can understand their expectations, preferences, and pain points. This enables you to refine prompts to deliver more satisfactory and user-friendly interactions, improving overall user satisfaction and engagement.

6. **Ethical Considerations:** Monitoring and improving prompts also play a role in addressing ethical considerations, such as transparency, privacy, and responsible AI practices. By regularly assessing prompts and their impact on the language model's outputs, you can identify and rectify any ethical concerns that may arise. This helps ensure that prompts promote responsible AI

behavior, respect user privacy, and maintain transparency in the model's decision-making processes.

In summary, continuous monitoring and improvement in prompt design and implementation are essential for adapting to changing user needs, addressing biases and fairness concerns, enhancing prompt effectiveness and performance, ensuring user satisfaction, and upholding ethical considerations. By actively monitoring, evaluating, and refining prompts, you can create a more robust and user-centric system that generates accurate, relevant, and responsible outputs.

Chapter 2.6: Addressing Ethical Concerns in Prompt Engineering

Identifying and mitigating biases in prompts and generated responses is an essential aspect of responsible prompt engineering. Here are some key strategies to address biases:

1. **Bias Detection and Evaluation:** Conduct thorough evaluations of prompts and generated responses to identify potential biases. Analyze the output for biases related to gender, race, religion, or other sensitive attributes. Use techniques such as manual review, human evaluation, or automated bias detection tools to identify biased patterns or language.
2. **Diverse and Representative Training Data:** Ensure that the training data used to develop language models and prompts is diverse and representative of different demographic groups and perspectives. Incorporate data from various sources and contexts to reduce the risk of bias. Consider including specific data augmentation techniques or pre-processing steps to increase the diversity of training data.
3. **Bias Mitigation Techniques:** Employ techniques to mitigate biases in prompts and generated responses. This may include adjusting the training data by removing or reweighting biased examples, augmenting the data with counterexamples, or applying debiasing algorithms during the training process. Carefully design prompt structures and formats to minimize the potential for biased responses.
4. **User Feedback and Iterative Improvement:** Encourage users to provide feedback on prompt-guided interactions and generated

responses. Establish feedback loops to capture user experiences and identify instances of bias. Actively incorporate user feedback into prompt refinement and model optimization processes to iteratively improve the system's performance and reduce biases over time.

5. **Collaboration with Domain Experts and Stakeholders:** Engage domain experts, ethicists, and stakeholders in the prompt engineering process. Seek their input and perspectives to identify potential biases and ethical concerns. Collaborative efforts can help uncover biases that may be overlooked and ensure a more comprehensive and inclusive approach to prompt design and evaluation.

6. **Regular Bias Audits and Reviews:** Conduct regular audits and reviews of prompt-guided systems to detect and address biases. Implement internal review processes or seek external audits to ensure independent scrutiny. Continuously monitor the system's performance, assess its impact on different user groups, and take corrective actions if biases are identified.

7. **Transparent Reporting and Documentation:** Provide transparent reporting on the prompt engineering process, including the steps taken to identify and mitigate biases. Document the considerations, techniques, and methodologies employed to address biases in prompts and generated responses. This documentation promotes transparency, accountability, and enables external scrutiny.

8. **Evaluation from Multiple Perspectives:** Consider diverse perspectives and feedback during the evaluation of prompt-generated responses. Solicit input from users, domain experts, and affected communities to assess the fairness and inclusivity of the system's outputs. Engaging multiple perspectives helps uncover biases that may be experienced differently by various user groups.

9. **Continuous Learning and Improvement:** Stay updated with research and best practices in bias mitigation techniques. Prompt engineering is an evolving field, and new methods and tools for bias detection and mitigation may emerge over time. Stay informed about the latest advancements and continuously improve prompt engineering practices to address biases effectively.

By implementing these strategies, prompt engineers can actively identify and mitigate biases in prompts and generated responses. This fosters

fairness, inclusivity, and responsible AI practices, ultimately leading to more equitable and unbiased interactions between users and AI systems.
 Ensuring fairness, inclusivity, and responsible AI practices in prompt design is of utmost importance to create ethical and unbiased language models. Here are some considerations to promote these principles in prompt design:

1. **Recognize Bias:** Be aware of potential biases in the prompts and their impact on the model's outputs. Bias can be introduced through the language, examples, or context provided in the prompts. Regularly evaluate prompts for biases related to gender, race, religion, and other sensitive attributes. Actively address and mitigate biases to ensure fairness and inclusivity in the model's responses.
2. **Diverse and Representative Examples:** Incorporate a diverse range of examples in prompts to ensure inclusivity and avoid reinforcing stereotypes or biases. Use data and examples that represent different demographics, cultures, and perspectives. This helps the model to learn from a more comprehensive and representative dataset, leading to fairer and more inclusive responses.
3. **Test for Fairness:** Perform fairness tests to evaluate how prompts may affect different groups or individuals. Assess the model's responses across various demographic categories to identify potential biases or disparities. If bias is detected, refine the prompts to mitigate these issues and ensure fair treatment for all users.
4. **Collaborate with Diverse Stakeholders:** Involve diverse stakeholders, including individuals from different backgrounds, communities, and domains, in the prompt design process. Seek input from experts, users, and affected communities to gain insights into potential biases or fairness concerns. Engaging diverse perspectives helps in addressing blind spots and ensures a more inclusive and responsible prompt design.
5. **Transparent Prompt Guidelines:** Provide clear and transparent guidelines for prompt design, specifying the expectations and requirements for generating unbiased and responsible responses. Clearly communicate the objectives and intended use cases of the language model to prompt designers, ensuring they understand the importance of fairness, inclusivity, and responsible AI practices.

6. **Regular Evaluation and Feedback:** Continuously evaluate prompts and the model's responses to assess their fairness, inclusivity, and adherence to responsible AI practices. Solicit feedback from users and domain experts to gain insights into any unintended biases or ethical concerns. Actively incorporate this feedback into prompt refinement processes.
7. **Addressing Sensitive Topics:** When designing prompts that involve sensitive topics, such as religion, politics, or social issues, exercise caution and ensure that the prompts are respectful, neutral, and do not promote harmful biases or discrimination. Consider providing guidelines to prompt designers on addressing sensitive topics responsibly.
8. **Ethical Review:** Establish an ethical review process for prompt design and implementation. This involves involving ethics committees or experts who can assess the ethical implications of the prompts, provide guidance, and ensure adherence to responsible AI practices.

Remember that ensuring fairness, inclusivity, and responsible AI practices in prompt design is an ongoing effort. Regularly revisit and refine prompts, consider user feedback, and stay updated with evolving societal norms and ethical guidelines. By proactively addressing biases, promoting inclusivity, and incorporating responsible AI practices, you can create language models that respect diversity, promote fairness, and contribute positively to society.
Ethical considerations related to data privacy, consent, and transparency are crucial when designing prompt-driven interactions. Here are some key points to consider:

1. **Data Privacy:** Ensure that user data is handled with utmost care and in compliance with relevant privacy regulations. Design prompts in a way that minimizes the collection and retention of personally identifiable information (PII) unless absolutely necessary. Implement robust data security measures to protect user data from unauthorized access or misuse.
2. **Informed Consent:** Obtain informed consent from users regarding the collection, use, and storage of their data. Clearly communicate the purpose of data collection and how it will be utilized in the prompt-driven interactions. Give users the ability to provide

explicit consent or opt-out if they are uncomfortable with sharing certain information.

3. **Transparent Data Handling:** Be transparent about how user data is used within the prompt-driven interactions. Clearly communicate what data is collected, why it is collected, and how it is utilized to enhance the user experience. Provide accessible privacy policies and terms of service that outline the data handling practices associated with the prompt-driven interactions.

4. **User Control and Preferences:** Empower users to have control over their data and the prompts they encounter. Provide options for users to customize their preferences, such as opting in or out of specific prompts or prompt types. Allow users to easily access and manage their data, including the ability to delete or update their information.

5. **Explain Prompt Functionality:** Clearly explain to users how prompts are utilized to guide the language model's responses. Inform users that their inputs, including prompts, may be used to train and improve the language model. Highlight that prompts are processed by the system and that their data is not shared or disclosed without their consent.

6. **Limit Scope and Context of Prompts:** When designing prompts, be mindful of the scope and context of the information being requested. Only ask for information that is directly relevant to the prompt-driven interactions and necessary for providing accurate and personalized responses. Minimize the collection of unnecessary or sensitive information.

7. **Regular Auditing and Compliance:** Regularly audit prompt-driven interactions and associated data handling practices to ensure compliance with ethical standards, privacy regulations, and internal policies. Stay updated with evolving legal and ethical guidelines related to data privacy and consent.

8. **User Education:** Educate users about the prompt-driven interaction process, the use of prompts, and the underlying language model's capabilities. Provide clear information about the limitations of the system and any potential risks or biases associated with the prompt-driven responses.

By incorporating these ethical considerations into the design and implementation of prompt-driven interactions, you can prioritize data privacy, obtain informed consent, ensure transparency, and empower users to have control over their data and the prompts they encounter. This fosters trust, respect user autonomy, and promotes responsible and ethical use of prompt-driven systems.

Chapter 2.7: Advanced Topics in Prompt Engineering

Exploring advanced techniques such as reinforcement learning for prompt optimization can offer valuable insights and improvements in prompt engineering. Here's an overview of how reinforcement learning can be applied to optimize prompts:

1. **Reinforcement Learning Basics:** Reinforcement learning (RL) is a machine learning technique that involves an agent interacting with an environment, learning to take actions that maximize a reward signal. In the context of prompt optimization, the language model acts as the RL agent, the prompt serves as the action, and the reward signal reflects the desired outcomes or performance metrics.

2. **Reward Design:** Designing an appropriate reward function is crucial in RL-based prompt optimization. The reward should reflect the desired objectives, such as generating accurate, diverse, or contextually relevant responses. The reward function can be defined based on evaluation metrics, user feedback, or other relevant criteria. It should incentivize prompts that lead to improved performance and discourage prompts that result in undesired behavior.

3. **Exploration and Exploitation:** RL algorithms balance exploration and exploitation to find optimal prompts. Initially, the model may explore a range of prompts to discover promising ones. Over time, it shifts towards exploiting the most effective prompts that maximize the reward. This iterative process allows the model to learn and refine its prompt selection strategy.

4. **Prompt Generation Strategies:** Reinforcement learning can be used to guide the generation of new prompts. The model can generate and evaluate a set of candidate prompts, selecting the most promising ones based on the reward signal. This approach allows for the exploration of a wide range of prompt possibilities

and can lead to the discovery of effective prompt structures or formats.

5. **Online Learning and Adaptive Prompts:** Reinforcement learning enables online learning, where the model can continually update its prompt selection based on real-time feedback. It can adapt to changing user preferences, evolving tasks, or dynamic environments. This flexibility allows for prompt optimization in dynamic scenarios where prompt effectiveness may vary over time.

6. **Reinforcement Learning Techniques:** Various RL algorithms and techniques can be applied to prompt optimization, such as Proximal Policy Optimization (PPO), Deep Q-Networks (DQN), or Policy Gradient methods. These techniques optimize the model's prompt selection policy by maximizing the expected reward signal.

7. **Balancing Trade-Offs:** Reinforcement learning can help navigate the trade-offs involved in prompt optimization. For example, it can balance the need for specific guidance with the desire for open-ended responses. By optimizing prompts with respect to various objectives and constraints, RL can help strike the right balance between controlled outputs and creative language generation.

While reinforcement learning offers promising avenues for prompt optimization, it requires careful design, appropriate reward shaping, and extensive experimentation. The complex nature of RL algorithms may necessitate large-scale training and computational resources. Nonetheless, by leveraging RL techniques, prompt optimization can be taken to a more advanced level, leading to improved performance, tailored responses, and enhanced user experiences in NLP tasks and applications.

Multimodal prompts refer to the integration of multiple modalities, such as text, images, audio, and video, to enhance the prompt-driven interactions and improve the output generated by language models. Here's an exploration of the benefits and challenges associated with multimodal prompts:

Benefits of Multimodal Prompts:

1. **Enhanced Contextual Understanding:** By combining multiple modalities, multimodal prompts provide richer and more

comprehensive context to the language model. This additional context can help the model better understand user intent, clarify ambiguous queries, and generate more accurate and contextually relevant responses.

2. **Improved User Experience:** Integrating different modalities allows users to express themselves in various ways beyond text alone. Users can provide visual cues, audio prompts, or multimedia references to enhance their communication with the language model. This can lead to a more engaging and interactive user experience.

3. **Enhanced Creativity and Expressiveness:** Multimodal prompts open up possibilities for creative expression by enabling users to convey their ideas, concepts, or emotions through a combination of text, images, audio, and video. This can be particularly beneficial for tasks that involve storytelling, content generation, or artistic endeavors.

4. **Better Handling of Ambiguity:** Ambiguity in natural language can be challenging for language models. By incorporating additional modalities, multimodal prompts can help disambiguate queries and provide clearer instructions or constraints to the model. This improves the chances of generating the desired output.

Challenges and Considerations:

1. **Data Availability and Integration:** Multimodal prompt engineering requires access to and integration of diverse datasets that include relevant text, images, audio, or video. Collecting, curating, and aligning multimodal data can be a complex and time-consuming process, requiring careful consideration of licensing, privacy, and data rights.

2. **Model Architecture and Training:** Existing language models are primarily designed for text inputs. Adapting these models to effectively process multimodal inputs requires architectural modifications and specialized training techniques. Developing multimodal models and training pipelines can be resource-intensive and computationally demanding.

3. **Intermodality Alignment and Fusion:** Integrating multiple modalities requires aligning and fusing information from different sources. Determining the optimal way to combine text, images, audio, and video inputs is an ongoing research challenge. Effective

fusion techniques need to be developed to leverage the strengths of each modality while mitigating potential biases or conflicts.

4. **Evaluation Metrics:** Evaluating the performance of multimodal prompts poses challenges in defining appropriate evaluation metrics. Traditional text-based metrics may not capture the full extent of the benefits and improvements achieved through multimodal prompt engineering. Developing comprehensive evaluation frameworks that consider various modalities is an active area of research.

5. **Accessibility and Inclusivity:** It's important to ensure that multimodal prompts are designed with accessibility and inclusivity in mind. Considerations should be given to individuals with visual or hearing impairments, providing alternative text descriptions, transcripts, or captioning to make the multimodal interactions accessible to a wider audience.

Despite these challenges, multimodal prompts hold great potential for advancing natural language processing applications. As research and technology progress, integrating text with other modalities will continue to enrich the prompt-driven interactions, enabling more nuanced and contextually aware language generation.

Personalized prompts involve tailoring the prompts used in prompt-driven interactions to the preferences, needs, and characteristics of individual users. Here's an exploration of the importance and benefits of personalized prompts:

1. **Enhanced User Experience:** Personalized prompts can significantly improve the user experience by aligning with users' specific preferences and needs. By adapting prompts to individual users, prompt-driven interactions can feel more intuitive, relevant, and engaging. This customization can lead to higher user satisfaction and increased engagement.

2. **Improved Task Relevance:** Personalized prompts enable a better alignment between the user's task requirements and the generated responses. By considering the user's specific context, goals, and preferences, prompts can be designed to address their unique needs, resulting in more accurate and actionable outputs.

3. **Contextual Understanding:** Personalized prompts take into account the user's historical interactions, preferences, and demographic information. This contextual understanding allows for more

nuanced and tailored prompts that consider the user's background, interests, and previous interactions. It can lead to more contextually relevant and personalized responses.

4. **Adaptability and Learning:** Personalized prompts can adapt to changes in user preferences over time. By continuously learning from user feedback and behavior, the prompts can be refined and adjusted to better meet the user's evolving needs. This adaptability helps in maintaining relevance and improving the overall user experience.

5. **Efficient Communication:** Personalized prompts can facilitate more efficient communication between the user and the language model. By leveraging insights about the user's preferred style, vocabulary, or tone, prompts can guide the model towards generating outputs that closely align with the user's communication style. This can save time and effort in clarifying or rephrasing prompts.

6. **Targeted Information Gathering:** Personalized prompts can be designed to elicit specific information from the user that is relevant to their preferences or requirements. By gathering targeted information, the prompts can guide the language model towards generating more precise and tailored responses.

7. **User Empowerment:** Personalized prompts put users in control of their prompt-driven interactions. By allowing users to customize or provide input for their prompts, they feel empowered and more involved in shaping the outcomes. This sense of ownership can contribute to a positive user experience and foster trust and satisfaction.

However, it's important to consider privacy and ethical implications when implementing personalized prompts. User data should be handled with care, ensuring compliance with privacy regulations, obtaining appropriate consent, and providing transparency about data usage and storage.

Personalized prompts offer significant potential for improving the effectiveness and user experience of prompt-driven interactions. By leveraging user-specific information, preferences, and historical data, prompts can be tailored to individual needs, resulting in more relevant, accurate, and satisfying responses.

Chapter 2.8: Future Directions and Challenges

Prompt engineering is an evolving field within natural language processing (NLP) with several emerging trends and potential advancements. Here are some notable trends and advancements to consider:

1. **Reinforcement Learning and Advanced Optimization Techniques:** Reinforcement learning (RL) techniques are being explored to optimize prompts and improve the performance of language models. RL algorithms enable models to learn from interactions and adapt prompt selection strategies over time. Advanced optimization techniques, such as evolutionary algorithms or Bayesian optimization, are also being investigated to automatically search for optimal prompts.
2. **Multimodal Prompts:** Integrating multiple modalities, such as text, images, audio, and video, in prompts is gaining attention. Multimodal prompts provide richer context and enable more expressive and interactive prompt-driven interactions. Research is focusing on effectively leveraging multimodal prompts to enhance the capabilities of language models and improve user experiences.
3. **Active Learning and Human-in-the-Loop Approaches:** Active learning techniques involve actively selecting informative prompts to query users, thereby reducing the amount of labeled data needed for training. Human-in-the-loop approaches, such as interactive prompt refinement, allow users to provide iterative feedback on prompt outputs, helping to refine and improve the prompts over time.
4. **Domain-Specific Prompt Engineering:** Customizing prompts for specific domains or verticals is becoming more prevalent. Domain-specific prompt engineering involves designing prompts that align with the language, jargon, or requirements of a particular domain, improving the model's performance in domain-specific tasks or applications.
5. **Explainable Prompt Engineering:** The interpretability and explainability of prompt-engineered models are gaining attention. Researchers are exploring techniques to provide insights into how prompts influence model behavior, enable interpretability of

prompt-driven outputs, and address concerns related to bias, fairness, and ethical considerations.
6. **Cross-Lingual and Multilingual Prompt Engineering:** Prompt engineering techniques are being extended to support cross-lingual and multilingual scenarios. This involves designing prompts that can effectively guide language models in generating responses in different languages or across language barriers, thereby enabling more inclusive and versatile applications.
7. **Benchmarking and Evaluation Metrics:** Efforts are being made to develop standardized benchmarks and evaluation metrics specifically for prompt engineering. This helps in comparing different prompt engineering approaches, understanding their strengths and limitations, and advancing the field through rigorous evaluation and benchmarking.
8. **User-Centric Prompt Design:** Prompt engineering is increasingly focusing on user-centric design principles. This involves actively involving users in the prompt design process, gathering user feedback, and considering user preferences and needs. User-centric prompt design aims to improve the user experience, ensure relevance, and meet user expectations in prompt-driven interactions.

As the field of prompt engineering continues to evolve, these trends and advancements are likely to shape the future direction of research and development. By harnessing the power of these emerging trends, prompt engineering can further enhance the capabilities and applications of language models, leading to more effective and tailored NLP systems. Prompt design, adaptation, and evaluation in natural language processing (NLP) pose several challenges and open research questions. Here are some key areas that researchers and practitioners are actively exploring:

1. **Bias and Fairness:** Designing prompts that mitigate bias and ensure fairness in language model outputs is a critical challenge. Open research questions include developing methods to identify and address biased prompts, evaluating the fairness of prompt-driven responses, and designing prompt adaptation techniques that promote fairness across diverse user groups.
2. **Generalization and Robustness:** Ensuring that prompts generalize well and produce robust outputs across different contexts and data distributions is an ongoing challenge. Research is focused on

understanding how prompt design and adaptation impact generalization capabilities and developing techniques to enhance the robustness of prompt-driven models in real-world scenarios.

3. **Explainability and Interpretability:** Prompt-driven models often lack transparency, making it challenging to interpret how prompts influence the model's responses. Research aims to develop explainable prompt engineering methods that provide insights into the reasoning behind model outputs and enable users to understand and trust the decision-making process.

4. **Data Efficiency:** Prompt adaptation techniques that can effectively leverage limited labeled data or actively query users for feedback to improve prompts are being explored. Efficient data collection strategies, active learning approaches, and transfer learning techniques are among the research directions aimed at improving prompt performance with limited data.

5. **Personalization and User Adaptation:** Customizing prompts for individual users and adapting prompts over time to align with changing user preferences and needs present challenges. Research questions include developing adaptive prompt engineering methods, designing personalized prompt strategies that respect privacy, and exploring user-centric approaches for prompt adaptation.

6. **Metrics for Evaluation:** Defining appropriate evaluation metrics for prompt-driven interactions is an ongoing challenge. Traditional NLP evaluation metrics may not fully capture the effectiveness of prompts in guiding language model responses. Developing robust and comprehensive evaluation metrics that consider various aspects of prompt quality and impact on model outputs is an active area of research.

7. **Human-AI Collaboration:** Exploring effective ways to combine human expertise with prompt engineering is crucial. Research focuses on developing collaborative prompt design frameworks that leverage human input, understanding how humans interact with prompt-driven systems, and identifying the optimal balance between human guidance and model autonomy.

8. **Ethical Considerations:** Ensuring ethical practices in prompt design, adaptation, and evaluation is essential. Open research questions involve addressing privacy concerns related to personalized prompts, understanding the potential ethical

implications of prompt-driven outputs, and developing guidelines and frameworks for responsible and ethical prompt engineering.

Addressing these challenges and answering these open research questions is key to advancing prompt design, adaptation, and evaluation in NLP. By addressing these areas, we can improve the effectiveness, fairness, and trustworthiness of prompt-driven interactions and unlock the full potential of language models in various applications.

As prompt engineering practices evolve and become more sophisticated, it is important to consider the ethical and societal implications associated with their use. Here are some key considerations:

1. **Bias and Fairness:** Prompt engineering can inadvertently introduce or amplify biases present in the training data or prompt design process. It is crucial to address biases in prompts to ensure fair and unbiased outputs. Ethical prompt engineering should strive to minimize bias and promote fairness across diverse user groups, taking into account factors such as race, gender, ethnicity, and other protected attributes.

2. **Misinformation and Disinformation:** Prompts that encourage or amplify misinformation or disinformation can have detrimental effects on society. Ethical prompt engineering practices should prioritize accuracy, fact-checking, and responsible information dissemination. Ensuring that prompts are aligned with credible sources and promoting critical thinking can help mitigate the spread of false information.

3. **Privacy and Data Protection:** Prompt engineering often involves the collection and analysis of user data. Respecting user privacy, obtaining appropriate consent, and protecting sensitive information are critical ethical considerations. Prompt engineering practices should adhere to privacy regulations and guidelines, implement secure data handling practices, and provide transparency regarding data usage.

4. **User Consent and Control:** Users should have control over their interactions with prompt-driven systems. Ethical prompt engineering should prioritize user consent, allowing users to opt-in or opt-out of prompt-driven interactions and providing clear mechanisms to control the prompts they receive. Users should also have the ability to modify or customize prompts according to their preferences.

5. **Transparency and Explainability:** Users should have insight into how prompts are generated and how they influence the language model's responses. Ethical prompt engineering practices should strive for transparency and provide explanations about the prompt design process and its impact on outputs. This fosters trust, accountability, and helps users make informed decisions.

6. **User Empowerment and Autonomy:** Prompt engineering should empower users and respect their autonomy. Users should have the freedom to provide feedback, challenge prompt-driven outputs, and influence the prompt design and adaptation process. Ethical practices should involve users as active participants, enabling them to shape their interactions and ensuring their voices are heard.

7. **Societal Impact:** Prompt engineering practices should consider the broader societal implications of the generated outputs. Language models have the potential to influence public discourse, shape opinions, and impact social norms. Ethical considerations involve proactively addressing harmful or offensive content, understanding the potential consequences of prompt-driven outputs, and promoting responsible AI practices.

8. **Accountability and Regulation:** As prompt engineering practices advance, it is important to establish clear accountability frameworks and regulations. Guidelines and standards should be developed to govern prompt engineering practices, addressing ethical concerns, ensuring transparency, and holding stakeholders accountable for the impacts of prompt-driven interactions.

Addressing these ethical and societal implications requires collaboration among researchers, practitioners, policymakers, and users. Ethical prompt engineering practices should prioritize fairness, accuracy, user consent, privacy, transparency, and user empowerment to ensure the responsible and beneficial use of prompt-driven systems in society.

By providing a comprehensive overview of the role of prompts in natural language processing, this book aims to equip readers with a deep understanding of prompt engineering techniques, ethical considerations, and the potential for further advancements in this exciting field.

Chapter 3: Designing Effective Prompts: Principles and Strategies for Maximum Impact

Designing effective prompts is crucial for achieving maximum impact in prompt-driven interactions. Here are some principles and strategies to consider:

1. **Clarity and Specificity:** Prompts should be clear and specific in communicating the desired task or user intent. Ambiguous or vague prompts may lead to confusion and inaccurate responses. Clearly define the expected input format, context, and desired output to guide the language model effectively.

2. **Relevance and Contextualization:** Prompts should be relevant to the task or application at hand and provide the necessary context for the language model to generate appropriate responses. Consider the specific domain, user requirements, and the intended use of the prompt-driven system. Tailor prompts to align with the user's needs and the context in which the system operates.

3. **Constraints and Guidelines:** Incorporate constraints or guidelines within the prompts to steer the language model's responses in a desired direction. Constraints can help ensure compliance with specific rules, policies, or ethical considerations. By providing explicit instructions or guidelines, you can guide the model's behavior and promote desired outcomes.

4. **Examples and Demonstrations:** Including examples and demonstrations within prompts can be helpful for guiding the language model's understanding and generating desired outputs. Showcasing sample inputs and corresponding desired responses can provide a clear illustration of the expected behavior. It helps the model grasp the nuances of the task and encourages more accurate and context-aware responses.

5. **Iterative Refinement:** Prompt design is often an iterative process. Start with an initial prompt and refine it based on feedback and evaluation results. Incorporate user feedback, identify areas of improvement, and iteratively update prompts to enhance their effectiveness. This iterative refinement process allows for continuous optimization and better alignment with user needs and expectations.

6. **User-Centric Design:** Consider the user's perspective when designing prompts. Understand the user's language, preferences, and context to create prompts that resonate with them. Engage

users in the prompt design process and solicit their input to ensure the prompts are tailored to their needs. User-centric prompt design fosters better user experiences and promotes user engagement.

7. **Evaluation and Feedback:** Regularly evaluate the effectiveness of prompts through user feedback, user testing, and performance metrics. Collect user feedback on prompt-driven interactions to identify areas for improvement and gain insights into the user experience. Use evaluation results to refine and optimize prompts for better performance and user satisfaction.

8. **Ethical Considerations:** Incorporate ethical considerations into prompt design. Avoid prompts that may lead to biased or discriminatory responses. Ensure prompts adhere to privacy regulations and respect user consent. Address ethical concerns related to fairness, inclusivity, and responsible AI practices in prompt design.

By applying these principles and strategies, you can design prompts that effectively guide language models and maximize the impact of prompt-driven interactions. Effective prompts improve the accuracy, relevance, and user experience, leading to more successful outcomes in various NLP tasks and applications.

Chapter 4: Unleashing Creativity: Prompt Variations and Techniques for Generating Diverse Outputs

Unleashing creativity in prompt-driven systems involves exploring various prompt variations and techniques to generate diverse outputs. Here are some strategies to consider:

1. **Open-Ended Prompts:** Instead of providing specific instructions or constraints, use open-ended prompts that allow the language model to freely explore and generate creative responses. Open-ended prompts encourage the model to think outside the box and unleash its creative potential.
2. **Contextual Prompts:** Introduce contextual cues within prompts to guide the language model's creativity. Provide relevant background information, setting descriptions, or character details to immerse the model in a specific context. This helps generate outputs that align with the given context and enhance the overall creativity of the responses.
3. **Creative Constraints:** Apply specific creative constraints within prompts to channel the language model's creativity in desired directions. For example, ask the model to generate responses in a particular style, mimic the voice of a famous author, or imagine alternative scenarios. These constraints provide a framework for creativity while still guiding the model's output.
4. **Visual and Multimodal Prompts:** Integrate visual stimuli, such as images or videos, with text prompts to inspire creative responses. Multimodal prompts can stimulate the model's imagination and result in more diverse and imaginative outputs. The combination of visual and textual cues provides a richer prompt context for creative generation.
5. **Random Seed Prompts:** Use random or unpredictable elements within prompts to trigger unexpected and novel responses. Incorporate random keywords, phrases, or concepts that can serve as creative triggers for the language model. This technique encourages the model to explore unconventional ideas and generate unique outputs.
6. **Prompt Expansion:** Expand or elaborate on the initial prompt by asking follow-up questions, providing additional details, or introducing new elements. This prompts the language model to

engage in a deeper exploration of the prompt and generate more imaginative and elaborate responses.

7. **Collaborative Prompt Generation:** Involve human collaboration in prompt generation. Engage users, writers, or other creative experts to contribute to the prompt design process. Their creative input can inspire the language model and lead to more imaginative outputs.

8. **Iterative Prompt Refinement:** Iterate on prompts by collecting user feedback and evaluating the creativity of the generated outputs. Refine prompts based on the feedback received, incorporating suggestions and insights to enhance the creative potential of the language model.

Remember to balance creativity with the desired objectives and constraints of the prompt-driven system. While encouraging diverse and imaginative outputs, it is important to ensure they align with the intended purpose, ethical considerations, and user expectations. Experimenting with different prompt variations and techniques can unlock the full creative potential of prompt-driven systems and lead to exciting and innovative results.

Chapter 5: Optimizing Prompt Structures: Crafting Clear, Specific, and Actionable Instructions

Optimizing prompt structures is crucial for crafting clear, specific, and actionable instructions that guide language models effectively. Here are some strategies to consider:

1. **Clear Language and Instructions:** Use clear and concise language in prompts to avoid ambiguity or confusion. Clearly communicate the desired task, expected format of the response, and any specific requirements. Ensure that the instructions are easy to understand for both the user and the language model.
2. **Specificity and Granularity:** Provide specific and detailed instructions to guide the language model's responses. Break down complex tasks into smaller, more manageable subtasks and provide instructions for each step. This helps the model understand and follow the prompt more accurately, leading to desired outputs.
3. **Actionable Prompts:** Craft prompts that encourage the language model to take specific actions or provide explicit information. Use action verbs and clear directives to prompt the model to perform a particular task or generate a specific type of response. Actionable prompts help guide the model's behavior and focus its attention on the desired outcome.
4. **Input/Output Format Guidance:** Specify the expected input and output formats in the prompts. For example, if you want the model to summarize a given text, explicitly state the desired length or format of the summary. If you require specific types of information in the response, provide guidelines on the structure or content. This helps the model produce responses that meet your requirements.
5. **Example-Based Instructions:** Include examples or demonstrations within prompts to illustrate the desired behavior. Show sample inputs and corresponding expected outputs to provide a clear reference for the language model. Examples help the model understand the prompt's intent and guide it towards generating responses that align with the provided examples.
6. **Task Clarification and Contextual Information:** Provide additional context or clarifications within prompts to help the language model better understand the task and its requirements. Include

relevant background information, specify the context or domain of the prompt, or explain any specific constraints. This additional information assists the model in generating more contextually appropriate responses.

7. **Iterative Refinement:** Iteratively refine prompts based on user feedback and evaluation results. Collect feedback from users or domain experts to identify areas for improvement and revise prompts accordingly. Conduct regular evaluations to assess the clarity, effectiveness, and user satisfaction with the prompts. Continuously refine and optimize the prompt structure based on these insights.

8. **User-Centric Approach:** Consider the user's perspective when designing prompts. Tailor the instructions to the target audience, taking into account their language proficiency, background knowledge, and familiarity with the task. Use user testing and feedback to validate the clarity and comprehensibility of the prompts from the user's point of view.

By following these strategies, you can optimize prompt structures to craft clear, specific, and actionable instructions that effectively guide language models. Clear instructions lead to better understanding, improved performance, and more accurate and relevant responses, ultimately enhancing the overall user experience.

Chapter 6: Leveraging Context: Incorporating Contextual Information in Prompts for Improved Performance

Leveraging contextual information in prompts is crucial for improving the performance of prompt-driven systems. By incorporating context, language models can generate more accurate, relevant, and context-aware responses. Here are some strategies to effectively leverage contextual information in prompts:

1. **Contextual Prompt Introductions:** Begin the prompt with contextual information that provides the necessary background for the language model. This can include relevant facts, previous statements, or a summary of the conversation history. By introducing context at the beginning of the prompt, you set the stage for the language model to generate responses that align with the given context.
2. **Task-specific Contextual Cues:** Include task-specific contextual cues within prompts to guide the language model's understanding and response generation. These cues can be specific keywords, phrases, or references that indicate the desired focus or direction for the model. For example, if the prompt is about movie recommendations, include information such as preferred genres or past movie ratings to guide the model's response.
3. **Conversation Flow and Dialogue Context:** In multi-turn conversations, incorporate the dialogue history or previous turns' context in the prompts. This helps the language model maintain continuity and coherence in its responses. By considering the conversation flow, the model can generate responses that are consistent with the ongoing dialogue and demonstrate a better understanding of the overall conversation.
4. **User-specific Information:** If available, incorporate user-specific information into prompts to personalize the responses. This can include user preferences, past interactions, or user profiles. By leveraging this information, the language model can generate responses that are tailored to the individual user's needs and preferences, enhancing the user experience.
5. **Temporal Context:** Consider the temporal aspect of the conversation or task when designing prompts. Incorporate relevant time-based information, such as recent events or changes, to provide up-to-date context for the language model. This helps the

model generate responses that are timely and aligned with the current situation.

6. **Domain-specific Context:** If the prompt is related to a specific domain or topic, ensure that the prompts include relevant domain-specific information. This can include industry-specific terminology, domain-specific constraints, or recent developments in the field. By incorporating domain-specific context, the language model can produce responses that are more accurate and relevant within that particular domain.

7. **Adaptive Prompting:** Continuously adapt prompts based on user feedback and ongoing interactions. Analyze user responses and adjust the prompts to incorporate new information or address any gaps in the model's understanding. Adaptive prompting ensures that the language model evolves with the changing context and user requirements.

8. **Evaluation and Iterative Refinement:** Regularly evaluate the effectiveness of prompts by measuring the performance of the language model and gathering user feedback. Assess the model's ability to understand and respond appropriately to contextual cues in the prompts. Use the evaluation results and user feedback to iteratively refine and optimize the prompts for improved performance.

By leveraging contextual information in prompts, you enhance the language model's ability to generate responses that are more accurate, relevant, and context-aware. This leads to improved performance, increased user satisfaction, and a more natural and engaging user experience in prompt-driven systems.

Chapter 7: Pitfalls and Challenges in Prompt Engineering: Common Mistakes and How to Avoid Them

Prompt engineering plays a crucial role in shaping the outputs of language models. However, there are several pitfalls and challenges that can arise in the process. Being aware of these common mistakes and knowing how to avoid them is essential for effective prompt engineering. Here are some pitfalls to watch out for and strategies to mitigate them:

1. **Ambiguous or Vague Prompts:** One common mistake is using prompts that are unclear or open to interpretation. Ambiguity in prompts can lead to inconsistent or irrelevant responses. To avoid this, ensure that prompts are specific, well-defined, and leave little room for ambiguity. Clearly communicate the desired task or outcome, provide explicit instructions, and avoid vague language.
2. **Bias in Prompts:** Unintentional bias can be introduced through prompts, resulting in biased or unfair responses from language models. It's important to carefully review and evaluate prompts for potential biases related to gender, race, ethnicity, or other sensitive attributes. Ensure that prompts are neutral, inclusive, and do not perpetuate harmful stereotypes. Incorporating diverse perspectives in prompt design can help mitigate bias.
3. **Insufficient Contextual Information:** Lack of context in prompts can lead to responses that are not aligned with the desired task or conversation. It's important to provide sufficient context to guide the language model's understanding and generate contextually appropriate responses. Include relevant information, previous dialogue history, or specific cues to help the model better comprehend the context.
4. **Overly Constraining Prompts:** While clear instructions are important, overly constraining prompts can restrict the creativity and flexibility of language models. Avoid being too prescriptive or limiting in prompts, as this can result in rigid or repetitive responses. Strike a balance between providing guidance and allowing room for the model's own creative interpretation.
5. **Inadequate Training Data:** Insufficient or low-quality training data can impact the performance of language models and their responsiveness to prompts. Ensure that the training data covers a wide range of topics, styles, and perspectives relevant to the desired task. Collect diverse and representative data to improve

the model's generalization and responsiveness to different prompts.

6. **Lack of Iterative Evaluation and Refinement:** Prompt engineering is an iterative process that requires continuous evaluation and refinement. Failing to evaluate the performance of prompts and iterate based on user feedback can hinder improvement. Regularly assess the quality of generated outputs, gather user feedback, and make necessary adjustments to prompts to enhance their effectiveness.

7. **Insensitive or Inappropriate Language:** Carelessly crafted prompts can result in language models generating inappropriate or offensive responses. It's important to be mindful of the potential for generating harmful or offensive content. Review prompts for sensitivity, ensure they adhere to ethical guidelines, and consider implementing content moderation mechanisms to filter out inappropriate responses.

To avoid these pitfalls, involve diverse stakeholders, such as domain experts, linguists, and users, in the prompt design process. Conduct thorough evaluations of both prompts and the generated outputs, soliciting user feedback to identify any issues or areas for improvement. Regularly update and refine prompts based on the insights gained from evaluations and user feedback.

By being aware of these pitfalls and proactively addressing them, you can enhance the effectiveness, fairness, and reliability of prompt engineering, leading to better outcomes and user experiences in language model interactions.

Chapter 8: Fine-tuning Models: Techniques and Best Practices for Tailoring Models to Specific Prompts

Fine-tuning models is a crucial step in prompt engineering to tailor them to specific prompts and tasks. It allows you to adjust and optimize the pre-trained language models for improved performance and better alignment with the desired outputs. Here are some techniques and best practices for fine-tuning models to specific prompts:

1. **Dataset Preparation:** Prepare a high-quality and diverse dataset that is representative of the target task. Curate data that covers a wide range of examples, including positive and negative cases, different variations of prompts, and relevant contextual information. Ensure that the dataset is properly labeled and balanced to avoid biases.

2. **Task-specific Architecture Modifications:** Consider making architecture modifications to the pre-trained model to align it with the specific task requirements. This can include adjusting the model's architecture layers, adding task-specific layers, or modifying attention mechanisms. Experiment with different architectural configurations to optimize the model's performance for the target task.

3. **Hyperparameter Tuning:** Fine-tuning involves adjusting hyperparameters that govern the training process. Optimize hyperparameters such as learning rate, batch size, weight decay, and dropout rates through systematic experimentation. Use techniques like grid search or Bayesian optimization to find the optimal combination of hyperparameters that yields the best performance.

4. **Transfer Learning:** Leverage transfer learning by initializing the model with pre-trained weights from a large-scale language model such as GPT-3 or BERT. This initialization helps the model capture general language understanding and improves the fine-tuning process. By starting from a pre-trained model, you can benefit from the knowledge and representations learned during the pre-training phase.

5. **Gradual Unfreezing:** Gradually unfreeze the layers of the model during fine-tuning to allow for better adaptation to the specific prompts. Start by freezing most layers and only update the last few layers initially. Then, progressively unfreeze and fine-tune earlier

layers of the model to capture task-specific patterns and avoid catastrophic forgetting.

6. **Regularization Techniques:** Apply regularization techniques to prevent overfitting and improve generalization. Common techniques include dropout, weight decay, and early stopping. Regularization helps the model learn more robust and generalizable representations by reducing the model's reliance on specific training examples.

7. **Domain Adaptation:** If the prompt is specific to a particular domain, consider incorporating domain adaptation techniques during fine-tuning. These techniques help the model adapt to the target domain by leveraging additional in-domain data or using domain-specific regularization methods. Domain adaptation can enhance the model's performance and alignment with the prompts in the target domain.

8. **Evaluation and Iterative Refinement:** Continuously evaluate the performance of the fine-tuned model on validation or test data. Measure metrics such as accuracy, precision, recall, or any domain-specific evaluation criteria. Analyze the model's outputs, solicit user feedback, and iteratively refine the fine-tuning process to address any shortcomings or areas of improvement.

9. **Data Augmentation:** Augment the training dataset by incorporating techniques such as data synthesis, paraphrasing, or back-translation. Data augmentation can help increase the diversity and variability of the training examples, leading to better generalization and improved performance on different prompt variations.

10. **Regular Monitoring and Retraining:** Prompt engineering is an ongoing process, and models should be regularly monitored and retrained as new data becomes available or the prompt requirements evolve. Continuously assess the model's performance, collect user feedback, and retrain the model to maintain its effectiveness over time.

By employing these techniques and following best practices, you can effectively fine-tune models to specific prompts, achieving better alignment with the desired outputs and optimizing performance for the target task.

Chapter 9: Evaluating Prompt Effectiveness: Metrics and Methods for Assessing Prompt Performance

Evaluating the effectiveness of prompts is crucial in prompt engineering to ensure that they are guiding language models to produce the desired outputs. Here are some metrics and methods that can be used to assess the performance of prompts:

1. **Task-specific Metrics:** Define task-specific metrics that measure the performance and success of the prompt in achieving the desired task outcome. For example, in text classification tasks, metrics like accuracy, precision, recall, and F1 score can be used. In text generation tasks, metrics like perplexity, BLEU score, or human evaluation ratings can be employed.
2. **Qualitative Evaluation:** Conduct qualitative evaluation by manually inspecting and assessing the generated outputs for prompt relevance, coherence, and overall quality. Expert human evaluators can provide subjective judgments and insights on the effectiveness of prompts in eliciting desired responses. Qualitative evaluation can also involve user feedback and user studies to gather insights on prompt performance.
3. **Comparison to Baselines:** Establish baseline models or existing approaches and compare the performance of prompts against them. Measure the improvements achieved with the prompt-guided model in terms of accuracy, quality, or other relevant metrics. A controlled experimental setup ensures fair comparisons and helps assess the added value of prompts.
4. **Human Evaluation:** Conduct human evaluation studies where evaluators, preferably domain experts or target users, assess the quality and relevance of generated outputs. Use evaluation criteria such as relevance to the prompt, fluency, coherence, and overall usefulness. Human evaluation can provide valuable insights into prompt effectiveness, especially in subjective or creative tasks.
5. **Diversity and Novelty:** Consider metrics that capture the diversity and novelty of the generated outputs. For example, metrics like distinct n-grams or unique responses can measure the diversity of responses. Novelty metrics can assess how different the generated outputs are from the training data, ensuring that prompts promote creative and unique responses.

6. **Prompt Impact Analysis:** Analyze the impact of different prompt variations or modifications on the model's outputs. Compare the performance and quality of generated responses across different prompts to understand the influence of prompt design choices. This analysis helps identify effective prompt strategies and informs prompt optimization.

7. **Consistency and Robustness:** Evaluate the consistency and robustness of prompt-guided models by measuring their performance across various input variations or slight modifications to the prompts. Assess how sensitive the models are to changes in prompt wording, format, or context. A robust prompt should consistently guide the model to produce reliable and consistent outputs.

8. **Adversarial Testing:** Conduct adversarial testing to assess the model's vulnerability to biased or manipulative prompts. Evaluate the model's responses to prompts designed to induce biased or harmful outputs. Adversarial testing helps identify potential ethical risks and informs prompt adjustments to mitigate biases or prevent misuse.

9. **Generalization Performance:** Assess how well the prompts enable the model to generalize to unseen or out-of-domain examples. Evaluate the performance of the model on a separate validation or test set that represents real-world scenarios. A prompt should enable the model to generalize well and produce desired outputs beyond the training data.

10. **Continuous Monitoring and User Feedback:** Implement systems for continuous monitoring of prompt-guided interactions, including user feedback mechanisms. Collect user feedback on the quality, relevance, and satisfaction with the generated responses. User feedback can provide valuable insights into prompt effectiveness and inform prompt improvements.

It's important to select appropriate evaluation methods and metrics based on the specific prompt engineering goals and the nature of the NLP task. Combine multiple evaluation approaches for a comprehensive assessment of prompt performance. Regularly evaluate and refine prompts based on evaluation results and user feedback to continuously improve their effectiveness in guiding language models.

Chapter 10: Advanced Prompt Engineering: Incorporating Multimodal Inputs and Domain-Specific Knowledge

Advanced prompt engineering techniques go beyond traditional text-based prompts and incorporate multimodal inputs and domain-specific knowledge to enhance the performance and capabilities of language models. Here are two key aspects of advanced prompt engineering:

1. **Multimodal Inputs:** Traditional prompts are typically text-based, but by incorporating other modalities such as images, audio, and video, we can provide richer context and enable more comprehensive and accurate responses from language models. Multimodal prompts allow language models to consider visual or auditory cues in addition to textual information when generating responses.

   - **Image-Based Prompts:** Use images as prompts to guide the language model's understanding and generation. Images can provide visual context, objects, scenes, or even specific instructions to generate more accurate and relevant responses. For example, an image of a cat can be used as a prompt to generate a description or answer questions about the cat.
   - **Audio-Based Prompts:** Incorporate audio inputs as prompts to capture speech or sound-related tasks. For example, an audio clip of a question can be used to generate a spoken response or transcribe the audio content into text. Audio-based prompts can be valuable in applications like voice assistants or transcription services.
   - **Video-Based Prompts:** Utilize video prompts to capture complex visual and temporal information. Video-based prompts enable language models to generate responses based on the visual content and temporal dynamics within the video. For example, a video clip of a cooking tutorial can be used as a prompt to generate step-by-step instructions or answer specific questions related to the video.

Integrating multimodal inputs requires advanced architectures and models that can effectively process and interpret different modalities. Techniques such as multimodal fusion, attention mechanisms, and pre-training on multimodal data can be employed to handle these inputs effectively.

2. **Domain-Specific Knowledge Incorporation:** Advanced prompt engineering also involves leveraging domain-specific knowledge to improve the performance and specificity of language models for particular domains or industries. By infusing domain-specific information into prompts, we can guide the models to generate more accurate and tailored responses.

- Domain-Specific Terminology: Use domain-specific terminology, keywords, or jargon in prompts to ensure that the language model understands and generates responses specific to a particular field. This helps in tasks like legal document generation, medical diagnoses, or technical support.

- Custom Knowledge Base: Integrate domain-specific knowledge bases or ontologies into prompts to provide structured information and constraints. This helps the language model generate responses that align with the domain-specific rules or guidelines. For instance, in legal applications, prompts can include references to specific laws, regulations, or legal precedents.

- Contextualizing Prompt with Domain Knowledge: Combine general prompts with domain-specific context to guide the model's responses. For example, when generating product descriptions, a prompt can include information about the product category, specifications, or customer preferences specific to the domain.

Incorporating domain-specific knowledge requires expertise in the target domain and effective integration of relevant data sources, knowledge graphs, or specialized lexicons. It helps language models generate responses that are more accurate, relevant, and aligned with the specific domain requirements.

By incorporating multimodal inputs and domain-specific knowledge, advanced prompt engineering techniques expand the capabilities of language models and enable them to generate more context-aware and domain-specific responses. These techniques enhance the versatility and performance of prompt-driven applications across various industries and use cases.

Chapter 11: Ethical Considerations in Prompt Engineering: Addressing Bias, Fairness, and Responsible AI

Ethical considerations play a crucial role in prompt engineering to ensure responsible and fair use of language models. Here are some key ethical considerations to address when designing prompts:

1. **Bias Awareness and Mitigation:** Language models are trained on large datasets, which may contain biases present in the data. These biases can lead to discriminatory or unfair outputs. It is important to identify and mitigate biases in prompts to ensure fairness and inclusivity. Consider techniques such as bias-corpus selection, data augmentation, or debiasing algorithms to reduce bias in prompts and improve the fairness of generated responses.
2. **Fairness in Output:** Evaluate the fairness of prompt-guided outputs across different demographic groups and sensitive attributes (e.g., race, gender, or religion). Assess whether the prompts lead to disparities in the quality or relevance of responses. If biases or unfairness are detected, take steps to rectify the issues by refining prompts or adjusting the training process.
3. **Transparency and Explainability**: Prompts should be designed in a way that promotes transparency and explainability of the model's behavior. Users should have a clear understanding of how prompts influence the model's responses. Provide transparency by disclosing the limitations, potential biases, and inherent uncertainties of the model. Document the prompt engineering process to facilitate audits and accountability.
4. **User Consent and Control:** Respect user autonomy by providing explicit consent mechanisms for prompt-driven interactions. Users should have the ability to opt-in or opt-out of prompt-based interactions. Allow users to customize prompts according to their preferences, ensuring that their values and sensitivities are taken into account.
5. **User Education and Guidelines:** Educate users about the nature and limitations of prompt-driven AI systems. Provide guidelines on appropriate use of prompts and potential ethical concerns. Foster user awareness of the impact of prompts on model behavior and encourage responsible usage.
6. **Continuous Monitoring and Evaluation:** Continuously monitor and evaluate prompt-guided outputs for potential biases or ethical

issues. Implement mechanisms to gather user feedback and reports of problematic outputs. Regularly update and refine prompts based on user feedback and emerging ethical guidelines.
7. **Ethical Review Processes:** Establish ethical review processes within organizations or research institutions to ensure that prompt engineering practices align with responsible AI principles. Conduct thorough assessments of potential ethical implications, including social, cultural, and privacy considerations, before deploying prompt-guided models in real-world applications.
8. **Collaboration and Accountability:** Engage diverse stakeholders, including ethicists, domain experts, and impacted communities, in the prompt engineering process. Foster collaboration and accountability to ensure that prompt designs consider a wide range of perspectives and ethical implications.

Addressing bias, fairness, and responsible AI practices in prompt engineering requires a multidisciplinary approach, involving not only technical expertise but also input from ethicists and stakeholders. By integrating ethical considerations throughout the prompt engineering lifecycle, we can mitigate potential harms, promote fairness, and ensure responsible AI practices in prompt-guided interactions.

Chapter 12: Future Trends and Directions in Prompt Engineering: Shaping the Path Ahead

Prompt engineering is a rapidly evolving field with ongoing advancements and future trends that shape its path ahead. Here are some key future trends and directions in prompt engineering:

1. **Adaptive and Dynamic Prompts:** The future of prompt engineering lies in developing prompts that can adapt and evolve based on user feedback and contextual cues. Adaptive prompts can adjust their structure, content, or style to improve the user experience and enhance the performance of language models. Dynamic prompts can leverage real-time user interactions to refine and optimize prompt-guided responses.

2. **Contextual Prompting:** Incorporating contextual information beyond the prompt itself will become increasingly important. Contextual prompts can include information about the user's preferences, location, or previous interactions to generate more personalized and relevant responses. Leveraging context-aware prompts will enhance the understanding of user intent and improve the overall user experience.

3. **Collaborative Prompting:** Prompt engineering can benefit from collaborative approaches, where users actively participate in co-creating prompts. By allowing users to provide feedback, suggest modifications, or contribute examples, prompt engineering can harness collective intelligence to improve prompt quality and effectiveness.

4. **Explainable Prompting:** As language models become more powerful and complex, there is a growing need for explainability in prompt engineering. Future trends will focus on developing techniques and methodologies to explain how prompts influence the model's responses. Explainable prompting will enhance transparency, trust, and accountability in prompt-guided AI systems.

5. **Domain-Specific Prompting:** Prompt engineering will continue to advance in domain-specific applications, catering to industries such as healthcare, finance, law, and more. Customizing prompts to specific domains will improve the accuracy and relevance of generated responses by incorporating domain knowledge, terminology, and constraints.

6. **Multilingual and Cross-Lingual Prompting:** Prompt engineering will expand to support multilingual and cross-lingual applications. Future trends will focus on developing prompts that effectively guide language models in generating responses in multiple languages. This will enable prompt-guided AI systems to serve diverse linguistic communities and facilitate cross-cultural interactions.

7. **Interdisciplinary Research:** Prompt engineering will increasingly draw insights and methodologies from diverse disciplines, including linguistics, cognitive science, psychology, and human-computer interaction. Interdisciplinary research collaborations will contribute to a deeper understanding of prompts' cognitive impact and enable the design of more effective prompts.

8. **Ethical and Responsible Prompt Engineering:** Future trends will emphasize the integration of ethical considerations and responsible AI practices in prompt engineering. This includes addressing biases, ensuring fairness, promoting transparency, and safeguarding user privacy and consent. Ethical frameworks and guidelines specific to prompt engineering will continue to evolve to navigate the ethical challenges in AI-driven interactions.

9. **Benchmarking and Evaluation:** The development of standardized benchmarks and evaluation metrics specific to prompt engineering will facilitate the comparison and assessment of different prompt engineering techniques and approaches. This will enable researchers and practitioners to measure the effectiveness and performance of prompt-guided AI systems accurately.

10. **User-Centric Design:** Prompt engineering will increasingly focus on user-centric design principles, considering user needs, preferences, and values. Future trends will emphasize user research, user feedback loops, and iterative design processes to ensure that prompts are intuitive, helpful, and aligned with user expectations.

By embracing these future trends and directions, prompt engineering will continue to advance, enabling more powerful, context-aware, and user-centric interactions with AI systems. It will contribute to the development of responsible and effective AI technologies that better serve the needs of individuals and society as a whole.

www.ingramcontent.com/pod-product-compliance
Lightning Source LLC
Chambersburg PA
CBHW070439220526
45466CB00004B/1734